Table of Electron Binding Energies (eV) for Electron Energy-Loss Spectra (EELS)

○ This table can be used to identify edges in electron energy-loss spectra

Notes:

1) This table is based on the values listed in Table B.1 (Appendix B).
2) Energies are listed up to 2000 eV for K-P edges; the lowest energy is given on the left.
3) In each series of edges (K-P), three or fewer of the lowest energies are listed from top to bottom.
4) Note that for a constituent element of a compound, the energy may be shifted by a few eV due to the chemical shift.

From ESCA (Almqvist and Wiksells Boktryckerz, Uppsala, 1967.)

	IA	IIA	IIIA	IVA	VA	VIA	VIIA	VIII	
1	**1 H** Hydrogen K 14								
2	**3 Li** Lithium K 55	**4 Be** Beryllium K 111							
3	**11 Na** Sodium M_1 1 $L_{2,3}$ 31, L_1 63 K 1072	**12 Mg** Magnesium M_1 2 $L_{2,3}$ 52, L_1 89 K 1305							
4	**19 K** Potassium $M_{2,3}$ 18, M_1 34 L_3 294, L_2 297, L_1 377	**20 Ca** Calcium $M_{4,5}$ 5, $M_{2,3}$ 26, M_1 44 L_3 347, L_2 350, L_1 438	**21 Sc** Scandium $M_{4,5}$ 7, $M_{2,3}$ 32, M_1 54 L_3 402, L_2 407, L_1 500	**22 Ti** Titanium $M_{4,5}$ 3, $M_{2,3}$ 34, M_1 59 L_3 455, L_2 461, L_1 564	**23 V** Vanadium $M_{4,5}$ 2, $M_{2,3}$ 38, M_1 66 L_3 513, L_2 520, L_1 628	**24 Cr** Chromium $M_{4,5}$ 2, $M_{2,3}$ 43, M_1 74 L_3 575, L_2 584, L_1 695	**25 Mn** Manganese $M_{4,5}$ 4, $M_{2,3}$ 49, M_1 84 L_3 641, L_2 652, L_1 769	**26 Fe** Iron $M_{4,5}$ 6, $M_{2,3}$ 56, M_1 95 L_3 710, L_2 723, L_1 846	**27 Co** Cobalt $M_{4,5}$ 3, $M_{2,3}$ 60, M_1 101 L_3 779, L_2 794, L_1 926
5	**37 Rb** Rubidium N_3 14, N_2 15, N_1 30 M_5 111, M_4 112, M_3 239 L_3 1806, L_2 1865	**38 Sr** Strontium $N_{2,3}$ 20, N_1 38 M_5 133, M_4 135, M_3 269 L_3 1941	**39 Y** Yttrium $N_{4,5}$ 3, $N_{2,3}$ 26, N_1 46 M_5 158, M_4 160, M_3 301	**40 Zr** Zirconium $N_{4,5}$ 3, $N_{2,3}$ 29, N_1 52 $M_{4,5}$ 180, M_4 183, M_3 331	**41 Nb** Niobium $N_{4,5}$ 4, $N_{2,3}$ 34, N_1 58 M_5 205, M_4 208, M_3 363	**42 Mo** Molybdenum $N_{4,5}$ 2, $N_{2,3}$ 35, N_1 62 M_5 227, M_4 230, M_3 393	**43 Tc** Technetium $N_{4,5}$ 2, $N_{2,3}$ 39, N_1 68 M_5 253, M_4 257, M_3 425	**44 Ru** Ruthenium $N_{4,5}$ 2, $N_{2,3}$ 43, N_1 75 M_5 279, M_4 284, M_3 461	**45 Rh** Rhodium $N_{4,5}$ 3, $N_{2,3}$ 48, N_1 81 M_5 307, M_4 312, M_3 496
6	**55 Cs** Cesium O_3 12, O_2 13, O_1 23 N_5 77, N_4 79, N_3 162 M_5 726, M_4 740, M_3 998	**56 Ba** Barium O_3 15, O_2 17, O_1 40 N_5 90, N_4 93, N_3 180 M_5 781, M_4 796, M_3 1063	57 La ~ 71 Lu	**72 Hf** Hafnium $O_{4,5}$ 7, O_3 31, O_2 38 N_7 18, N_6 19, N_5 214 M_5 1662, M_4 1716	**73 Ta** Tantalum $O_{4,5}$ 6, O_3 37, O_2 45 N_7 25, N_6 27, N_5 230 M_5 1735, M_4 1793	**74 W** Tungsten $O_{4,5}$ 6, O_3 37, O_2 47 N_7 34, N_6 37, N_5 246 M_5 1810, M_4 1872	**75 Re** Rhenium $O_{4,5}$ 4, O_3 35, O_2 46 N_7 45, N_6 47, N_5 260 M_5 1883, M_4 1949	**76 Os** Osmium $O_{4,5}$ 0, O_3 46, O_2 58 N_7 50, N_6 52, N_5 273 M_5 1960	**77 Ir** Iridium $O_{4,5}$ 4, O_3 51, O_2 63 N_7 60, N_5 63, N_5 295
7	**87 Fr** Francium $P_{2,3}$ 15, P_1 34 $O_{4,5}$ 58, O_3 140, O_2 182 $N_{6,7}$ 268, N_5 577, N_4 603	**88 Ra** Radium $P_{2,3}$ 19, P_1 44 $O_{4,5}$ 68, O_3 153, O_2 200 $N_{6,7}$ 299, N_5 603, N_4 636	89 Ac ~ 103 Lr						

57 La Lanthanum	58 Ce Cerium	59 Pr Praseodymium	60 Nd Neodymium	61 Pm Promethium	62 Sm Samarium
$O_{2,3}$ 15, O_1 33 $N_{4,5}$ 99, N_3 192, N_2 206 M_5 832, M_4 849, M_3 1124	$O_{2,3}$ 20, O_1 38 $N_{6,7}$ 1, $N_{4,5}$ 111, N_3 208 M_5 884, M_4 902, M_3 1186	$O_{2,3}$ 23, O_1 38 $N_{6,7}$ 2, $N_{4,5}$ 114, N_3 218 M_5 931, M_4 951, M_3 1243	$O_{2,3}$ 22, O_1 38 $N_{6,7}$ 2, $N_{4,5}$ 118, N_3 225 M_5 978, M_4 1000, M_3 1298	$O_{2,3}$ 22, O_1 38 $N_{6,7}$ 4, $N_{4,5}$ 121, N_3 237 M_5 1027, M_4 1052, M_3 1357	$O_{2,3}$ 22, O_1 39 $N_{6,7}$ 7, $N_{4,5}$ 130, N_3 249 M_5 1081, M_4 1107, M_3 1421

89 Ac Actinium	90 Th Thorium	91 Pa Protactinium	92 U Uranium	93 Np Neptunium	94 Pu Plutonium
$O_{4,5}$ 80, O_3 167, O_2 215 $N_{6,7}$ 319, N_5 639, N_4 675	$P_{4,5}$ 2, P_3 43, P_2 49 O_5 88, O_4 95, O_3 182 N_7 335, N_6 344, N_5 677	$O_{5,6}$ 94, $O_{2,3}$ 223, O_1 310 N_7 360, N_6 371, N_5 708	$P_{4,5}$ 4, P_3 33, P_2 43 O_5 96, O_4 105, O_3 195 N_7 381, N_6 392, N_5 738	O_5 101, O_4 109, O_3 206 N_7 404, N_6 415, N_5 773	O_5 105, O_4 116, O_3 212 $N_{6,7}$ 422, N_5 801, N_4 849

	IB	IIB	IIIB	IVB	VB	VIB	VIIB	0
								2 He Helium K 25
Symbol → 8 O (Atomic number → 8) Oxygen ← Name $L_{2,3}$ 7, K 532 ← Edge, Binding energy (eV) L_1 24			5 B Boron $L_{2,3}$ 5, K 188	6 C Carbon $L_{2,3}$ 7, K 284	7 N Nitrogen $L_{2,3}$ 9, K 399	8 O Oxygen $L_{2,3}$ 7, K 532 L_1 24	9 F Fluorine $L_{2,3}$ 9, K 686 L_1 31	10 Ne Neon $L_{2,3}$ 18, K 867 L_1 45
			13 Al Aluminum M_1 1, L_3 73, K 1560 L_2 74 L_1 118	14 Si Silicon $M_{2,3}$ 3, L_3 99, K 1839 M_1 8, L_2 100 L_1 149	15 P Phosphorus $M_{2,3}$ 10, L_3 135 M_1 16, L_2 136 L_1 189	16 S Sulfur $M_{2,3}$ 8, L_3 164 M_1 16, L_2 165 L_1 229	17 Cl Chorine $M_{2,3}$ 7, L_3 200 M_1 18, L_2 202 L_1 270	18 Ar Argon $M_{2,3}$ 12, L_3 245 M_1 25, L_2 247 L_1 320
28 Ni Nickel $M_{4,5}$ 4, L_3 855 $M_{2,3}$ 68, L_2 872 M_1 112, L_1 1008	29 Cu Copper $M_{4,5}$ 2, L_3 931 $M_{2,3}$ 74, L_2 951 M_1 120, L_1 1096	30 Zn Zinc $M_{4,5}$ 9, L_3 1021 $M_{2,3}$ 87, L_2 1044 M_1 137, L_1 1194	31 Ga Gallium $N_{2,3}$ 1, $M_{4,5}$ 18, L_3 1116 M_3 103, L_2 1143 M_2 107, L_1 1298	32 Ge Germanium $N_{2,3}$ 3, $M_{4,5}$ 29, L_3 1217 M_3 122, L_2 1249 M_2 129, L_1 1414	33 As Arsenic $N_{2,3}$ 3, $M_{4,5}$ 41, L_3 1323 M_3 141, L_2 1359 M_2 147, L_1 1527	34 Se Selenium $N_{2,3}$ 6, $M_{4,5}$ 57, L_3 1436 M_3 162, L_2 1476 M_2 168, L_1 1654	35 Br Bromine $N_{2,3}$ 5, M_5 69, L_3 1551 N_1 27, M_4 70, L_2 1597 M_3 182, L_1 1783	36 Kr Krypton $N_{2,3}$ 11, $M_{4,5}$ 89, L_3 1675 N_1 24, M_3 214, L_2 1727 M_2 223, L_1 1921
46 Pd Palladium $N_{4,5}$ 1, M_5 335 $N_{2,3}$ 51, M_4 340 N_1 86, M_3 531	47 Ag Silver $N_{4,5}$ 3, M_5 367 N_3 56, M_4 373 N_2 62, M_3 571	48 Cd Cadmium $O_{2,3}$ 2, $N_{4,5}$ 9, M_5 404 $N_{2,3}$ 67, M_4 411 N_1 108, M_3 617	49 In Indium $O_{2,3}$ 1, $N_{4,5}$ 16, M_5 443 $N_{2,3}$ 77, M_4 451 N_1 122, M_3 664	50 Sn Tin $O_{2,3}$ 1, $N_{4,5}$ 24, M_5 485 O_1 1, $N_{2,3}$ 89, M_4 494 N_1 137, M_3 715	51 Sb Antimony $O_{2,3}$ 2, $N_{4,5}$ 32, M_5 528 O_1 7, $N_{2,3}$ 99, M_4 537 N_1 152, M_3 766	52 Te Tellurium $O_{2,3}$ 2, $N_{4,5}$ 40, M_5 572 O_1 12, $N_{2,3}$ 110, M_4 582 N_1 168, M_3 819	53 I Iodine $O_{2,3}$ 3, $N_{4,5}$ 50, M_5 620 O_1 14, $N_{2,3}$ 123, M_4 631 N_1 186, M_3 875	54 Xe Xenon $O_{2,3}$ 7, $N_{4,5}$ 63, M_5 672 O_1 18, $N_{2,3}$ 147, M_4 685 N_1 208, M_3 937
78 Pt Platinum $O_{4,5}$ 2, N_7 70 O_3 51, N_6 74 O_2 66, N_5 314	79 Au Gold $O_{4,5}$ 3, N_7 83 O_3 54, N_4 87 O_2 72, N_5 334	80 Hg Mercury $O_{4,5}$ 7, N_7 99 O_3 58, N_6 103 O_2 81, N_5 360	81 Tl Thallium O_5 13, N_7 118 O_4 16, N_6 122 O_3 76, N_5 386	82 Pb Lead $P_{2,3}$ 1, O_5 20, N_7 138 P_1 3, O_4 22, N_6 143 O_3 86, N_5 413	83 Bi Bismuth $P_{2,3}$ 3, O_5 25, N_7 158 P_1 8, O_4 27, N_6 163 O_3 93, N_5 440	84 Po Polonium $P_{2,3}$ 5, $O_{4,5}$ 31, $N_{6,7}$ 184 P_1 12, O_3 104, N_5 473 O_2 132, N_4 500	85 At Astatine $P_{2,3}$ 8, $O_{4,5}$ 40, $N_{6,7}$ 210 P_1 18, O_3 115, N_5 507 O_2 148, N_4 533	86 Rn Radon $P_{2,3}$ 11, $O_{4,5}$ 48, $N_{6,7}$ 238 P_1 26, O_3 127, N_5 541 O_2 164, N_4 567
63 Eu Europium $O_{2,3}$ 22, $N_{6,7}$ 0, M_5 1131 O_1 32, $N_{4,5}$ 134, M_4 1161 N_3 257, M_3 1481	64 Gd Gadolinium $O_{2,3}$ 21, $N_{6,7}$ 0, M_5 1186 O_1 36, $N_{4,5}$ 141, M_4 1218 N_3 271, M_3 1544	65 Tb Terbium $O_{2,3}$ 26, $N_{6,7}$ 3, M_5 1242 O_1 40, $N_{4,5}$ 148, M_4 1276 N_3 286, M_3 1612	66 Dy Dysprosium $O_{2,3}$ 26, $N_{6,7}$ 4, M_5 1295 O_1 63, $N_{4,5}$ 154, M_4 1332 N_3 293, M_3 1676	67 Ho Holmium $O_{2,3}$ 20, $N_{6,7}$ 4, M_5 1351 O_1 51, $N_{4,5}$ 161, M_4 1391 N_3 306, M_3 1741	68 Er Erbium $O_{2,3}$ 29, $N_{6,7}$ 4, M_5 1409 O_1 60, N_5 168, M_4 1453 N_4 177, M_3 1812	69 Tm Thulium $O_{2,3}$ 32, $N_{6,7}$ 5, M_5 1468 O_1 53, $N_{4,5}$ 180, N_4 1515 N_3 337, M_3 1885	70 Yb Ytterbium $O_{2,3}$ 23, $N_{6,7}$ 6, M_5 1527 O_1 53, N_5 184, M_4 1576 N_4 197, M_3 1949	71 Lu Lutetium $O_{4,5}$ 5, $N_{6,7}$ 7, N_5 1589 $O_{2,3}$ 28, N_5 195, N_4 1640 O_1 57, N_4 205
95 Am Americium O_5 103, $N_{6,7}$ 440 O_4 116, N_5 828 O_3 220, N_4 879	96 Cm Curium	97 Bk Berkelium	98 Cf Californium	99 Es Einsteinium	100 Fm Fermium	101 Md Mendelevium	102 No Nobelium	103 Lr Lawrencium

Springer Japan KK

D. Shindo · T. Oikawa

Analytical Electron Microscopy for Materials Science

With 180 Figures

Daisuke Shindo, Dr.
Professor
Institute of Multidisciplinary Research for Advanced Materials
Tohoku University
2-2-1 Katahira, Sendai, Miyagi 980-8577, Japan

Tetsuo Oikawa, Dr.
Chief Researcher
Application and Research Center, JEOL Ltd.
3-1-2 Musashino, Akishima, Tokyo 196-8558, Japan

Front cover
Upper: Schematic illustration of energy dispersion of omega-type energy filter.
Lower: Electron diffraction patterns of $Ti_{50}Ni_{48}Fe_2$ without (*top*) and with (*bottom*) the energy filter. With the use of the energy filter, weak diffuse scattering showing the precursor phenomena to the R-phase transformation is clearly revealed between the fundamental reflections such as 011 and 110 reflections. The *insets* are intensity profiles. (See Sect. 3.6.3.2)

Back cover
Upper: Lorentz microscope image of rapidly quenched $Fe_{73.5}Cu_1Nb_3Si_{13.5}B_9$ showing magnetic domain walls as white and black bands.
Middle: Electron hologram of $Fe_{73.5}Cu_1Nb_3Si_{13.5}B_9$ observed with the use of a thermal FEG and a biprism.
Lower: Reconstructed phase image obtained from the electron hologram by Fourier transform, showing the lines of magnetic flux (*arrows*). (See Sect. 5.3.2.2)

Library of Congress Cataloging-in-Publication Data applied for.

ISBN 978-4-431-70336-5 ISBN 978-4-431-66988-3 (eBook)
DOI 10.1007/978-4-431-66988-3

This English translation is based on the Japanese original, D. Shindo, T. Oikawa; Analytical Electron Microscopy for Materials Analysis
Published by Kyoritsu Shuppan

Printed on acid-free paper

Originally published by Spinger-Verlag Tokyo in 2002.

Typesetting: SNP Best-set Typesetter Ltd., Hong Kong

SPIN: 10868581

Preface

To develop the advanced materials that will support new technology in the twenty-first century, it is essential to have a detailed knowledge of the structures of those materials. In fabrication of multilayered films and composite materials, for example, the quantitative information about structure and composition on a nanometer scale is indispensable to understand their properties. Nowadays, analytical transmission electron microscopy has attracted much attention as one of the best experimental methods owing to its superior performance of high-resolution imaging (~0.1 nm) and nanoprobe analysis (~1 nm in diameter). At the same time, it is quite true that the higher the performance of analytical electron microscopes becomes, the more thorough the knowledge about instrumentation hardware is required to be in order to give full play to analytical transmission electron microscopy.

Taking these facts into account, the authors planned a book that explains both the hardware and the software of current analytical transmission electron microscopes. Previously, in collaboration with Professor K. Hiraga, one of the authors (D.S.) published *High-Resolution Electron Microscopy for Materials Science*,[1] in which the optimum imaging conditions for obtaining fine high-resolution electron microscope images were discussed along with the appropriate techniques of analyzing those images. In contrast to that book, the aim of the present one is to explain the principles of transmission electron microscopes and their analytical techniques. Materials science forms the background of both books, however, and the term appears in both titles.

To present the content in an easily understandable way, many schematic illustrations and experimental data are included in the book. In addition, the principles of electron energy-loss spectroscopy (EELS) and energy dispersive X-ray spectroscopy (EDS) are clearly explained.

Chapter 1 notes the interactions between incident electrons and solids, which form the basis of analytical electron microscopy and its application. Basic parameters such as the scattering cross section and the mean free path, which are necessary for analytical electron microscopy, are also explained. In Chapter 2, the principles and performance of the hardware of analytical electron microscopes, such as field emission guns and electron lenses, are presented. The techniques for establishing optimum experimental conditions are explained. In Chapter 3, the basis and application of EELS are noted. In addition to the understanding of electron energy-loss spectra corresponding to various inelastic electron-scattering processes, the principles and application of the energy filtering method are also elucidated. Chapter 4 sets forth the principles and application of EDS, which has been used as the most fundamental analytical method in analytical electron microscopy. Some precautions are noted regarding quantitative compositional analysis. Also explained in

[1] Springer-Verlag, Tokyo, 1998

some detail is Atom Location CHanneling Enhanced MIcroanalysis (ALCHEMI), which combines EDS and the electron diffraction effect and is useful for locating impurity atoms. Finally, in Chapter 5, other analytical techniques of analytical transmission electron microscopes, such as convergent beam electron diffraction, electron holography, and scanning electron microscopy, are presented along with explanations of typical methods of specimen preparation.

The authors are grateful to many researchers at Tohoku University and JEOL Ltd. for useful discussions on analytical electron microscopy. For obtaining the analytical electron microscope data presented in this book, the collaboration of the authors' colleagues Professor K. Hiraga, Dr. A. Taniyama, Dr. Y. Murakami, Dr. J. Yang, Dr. Y. Ikematsu, Dr. Y.S. Lee, Dr. Z. Liu, Dr. M. Kawasaki, Dr. C.W. Lee, Mr. Y.G. Park, Mr. Y. Aoyama, and Mr. J.H. Yoo was invaluable. Special acknowledgments go to Chairman T. Etoh and President Y. Harada, JEOL Ltd., for their interest and encouragement.

D. Shindo and T. Oikawa
Sendai, Japan
July 2002

Contents

Notes

1. Basic Principles of Analytical Electron Microscopy

Before going into a detailed explanation of the hardware of transmission electron microscopes and analytical methods, it is necessary to understand some fundamental aspects. These areas include the interactions between incident electrons and materials, the basic principles of analytical electron microscopy, and the processing of analytical data.

1.1 Interaction Between Electrons and Materials

1.1.1 Scattering of Electrons

In the center of a *transmission electron microscope* column a specimen is illuminated with high-energy electrons, as shown in Fig. 1.1. There may be various interactions between this specimen and the incident electrons. When the specimen is extremely thin, many electrons penetrate the specimen without interactions, and these electrons are called *transmitted electrons*. The rest of the electrons interact more or less with the specimen; and the probability of the interactions increases with the increase in specimen thickness. Electron scattering caused by the specimen can be classified into two groups: *elastic scattering* and *inelastic scattering*. In the elastic scattering the direction of the scattered electrons changes, but their velocity (or energy) does not. *Diffracted electrons* and *back-scattered electrons* belong to the elastic scattering category. The back-scattering process is notable for lower energy electrons, but the probability of back-scattering is rather low in conventional transmission electron microscopes. On the other hand, all electrons suffering a change of their velocity (or energy) belong to the inelastic scattering category. Whereas imaging modes such as the *bright-field method*, *dark-field method*, and *high-resolution electron microscopy* [1] mainly utilize elastically scattered electrons, *analytical electron microscopy* utilizes inelastically scattered electrons. There are various inelastic scattering processes, with the main origins of the inelastic scattering processes as follows.

1. Lattice vibration (phonon excitation)
2. Collective excitation of valence electrons (plasmon excitation)
3. Interband transition
4. Inner-shell excitation (core excitation)
5. Excitation of free electrons (excitation of secondary electrons)
6. Bremsstrahlung (emission of continuous X-rays)

The specific energy loss for each excitation process may be found in Section 3.1. Spectroscopy of inelastically scattered electrons, taking into account these scattering processes, is called *electron energy-loss spectroscopy* (EELS). Spectroscopy of characteristic X-rays resulting from inner-shell excitation is *energy dispersive X-ray spectroscopy* (EDS or EDX, EDXS). EELS and EDS are the two methods most popularly used in analytical electron microscopy. To help the reader understand analytical electron microscopy, the basic principles of EELS and EDS are presented in Section 1.2.1, and the details of these methods and their application are described in Chapters 3 and 4.

1.1.2 Fundamental Quantities Characterizing Electron Scattering

Species and probability of electron scattering in materials depend on constituent elements, incident electron energy, and so on. The characteristics of electron scattering can be described in several fundamental areas.

1.1.2.1 Scattering Cross Section

Scattering processes can be expressed quantitatively with the scattering cross section σ. The scattering cross section, indicating the probability of the scattering event, is given by the following equation.

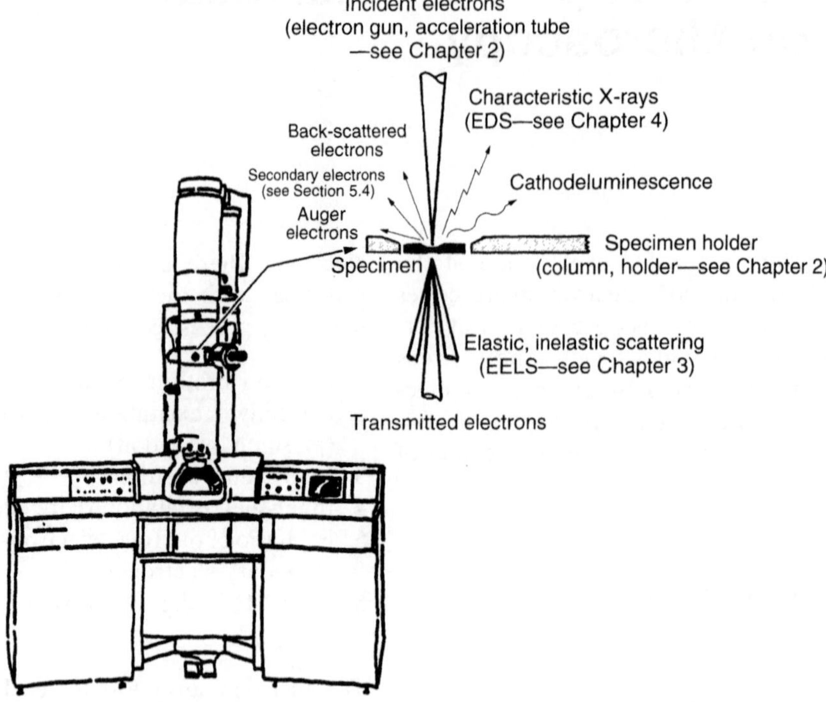

Fig. 1.1. Interaction between the incident electrons and a specimen in a transmission electron microscope (*TEM*). *EDS*, energy dispersion x-ray spectroscopy; *EELS*, electron energy-loss spectroscopy

$$\sigma = \frac{N}{n_m n_e} \tag{1.1}$$

where N (events/cm^3) is the number of specific scattering events in the unit volume, such as elastic scattering, inner-shell excitation and so on; n_m is the number of atoms in the unit volume; and n_e is the number of incident electrons in the unit area. The numbers of events, atoms, and electrons are dimensionless, and the cross section has the dimension of square centimeters. In general, the cross section for each scattering process can be theoretically given with various physical parameters, such as the scattering angle and energy loss. The scattering cross section including all the scattering process, which is theoretically given with the physical parameters, including the scattering angle and energy loss, is called the *total scattering process*. In contrast, the scattering cross section defined under limited conditions, such as for the specific scattering angle range and limited energy loss range, is called the *partial scattering cross section*.

Elastic Scattering Cross Section. The cross section concerned with elastic scattering is called the *elastic scattering cross section*. When the incident electrons pass near the nucleus of an atom, their trajectories curve owing the coulomb field of the nucleus. This event is called *nucleus scattering* or *Rutherford scattering*. The cross section of Rutherford scattering, taking into account the relativistic effect and screening of the nucleus due to the inner-shell electrons, was represented by Wentzel [2] and Mott and Massey [3] as follows.

$$\frac{d\sigma}{d\Omega} = \frac{Z^2 e^4}{16E^2}\{\sin^2(\Theta/2) + (\Theta_0^2/4)\}^{-2}$$
$$\{1 - \beta_r^2 \sin^2(\Theta/2) + \pi\alpha\beta_r[\sin(\Theta/2) - \sin^2(\Theta/2)]\} \tag{1.2}$$

where Ω is a solid angle. It is given as

$$d\Omega = 2\pi \sin\Theta d\Theta \tag{1.3}$$

Also, Θ ($0 \le \Theta \le \pi$) is the scattering angle and is related to the diffraction angle θ in Eq. 2.5 as follows.

$$\Theta = 2\theta \tag{1.4}$$

Z is the atomic number; E is the incident electron energy in kiloelectron volts (keV); Θ_0 is the screening parameter; and $\beta_r = v/c$ is given by

$$\beta_r = \left\{1 - [1 + (E/511)]^{-2}\right\}^{1/2} \tag{1.5}$$

(see also Appendix 1). In Eq. (1.5), 511 (keV) is the energy corresponding to the rest mass of the electron.

The expression $d\sigma/d\Omega$, the cross section for the element of solid angle, is called the *differential elastic scattering cross section*. Because the screening parameter Θ_0 [4] and the constant α [5] in Eq. 1.2 depend on the atomic number and the incident electron energy, the differential elastic scattering cross section is not simply proportional to $(Z/E)^2$; but in general it is known that the differential elastic scattering cross section increases with an increase in atomic number or a decrease in accelerating voltage.

Note that the atomic scattering factor, used for analyzing the diffraction intensity and evaluated on the basis of wave function, is related to the differential elastic scattering cross section as follows.

$$\frac{d\sigma}{d\Omega} = |f|^2 \tag{1.6}$$

Inelastic Scattering Cross Section. The cross section concerned with inelastic scattering is called the *inelastic scattering cross section*. There are many inelastic scattering processes, so the cross section is defined for each inelastic scattering process. The following is the differential inelastic scattering cross section for *plasmon excitation*.

$$\frac{d\sigma(\Theta)}{d\Omega} = \frac{1}{2\pi a_0} \frac{\Theta_p}{\Theta^2 + \Theta^2{}_p} \tag{1.7}$$

where a_0 is the Bohr radius (0.0529 nm), and Θ_p equals $\Delta E_p/2E$ (E_p is the plasmon excitation energy). ΔE_p ranges from several electron volts to 30 eV, being much smaller than the incident electron energy; hence Θ_p has a small value in Eq. 1.7. Thus, the inelastic scattering cross section for plasmon excitation decreases drastically with the increase in the scattering angle Θ.

The following equation corresponds to the inelastic scattering cross section for *inner-shell excitation*.

$$\sigma = \frac{\pi e^4 b_s n_s}{\left(\frac{m_0 v^2}{2}\right) E_c} \left\{ \log\left[c_s \left(\frac{m_0 v^2}{2}\right) \Big/ E_c \right] - \log\left(1 - \beta_r^2\right) - \beta_r^2 \right\} \tag{1.8}$$

where E_c is the energy for ionization; n_s is the number of electrons in the inner shell; m_0 is the rest mass of the electron; and v and e are the velocity of incident electrons and the elementary electric charge, respectively. The b_s and c_s values depend on the species of inner shells, such as the K shell, L shell, and so on. Thus the inelastic scattering cross section is known to be largely dependent on b_s and c_s, and these constants have been evaluated on the basis of experimental data [6, 7].

1.1.2.2 Mean Free Path

The average distance (λ_s) that the electron travels between the scattering events is called the mean free path. The following equation indicates the relation between the mean free path and the scattering cross section (σ).

$$\lambda = \frac{A}{\sigma N_0 \rho} (\text{cm}) \tag{1.9}$$

where N_0 and A are Avogadro's number and the atomic mass, respectively; and ρ is the density. Thus, the mean free path for the specific scattering process i can be obtained by inserting the scattering cross section σ_i into Eq. 1.9.

When an electron is scattered once in a specimen, so the mean free path is smaller than the thickness of the specimen, it is called *single scattering*; when the electron is scattered twice to several times inside the specimen, it is called *plural scattering*; and when the electron scatterings occur more than several times, it is called *multiple scattering*.

1.1.2.3 Beam Broadening

When the small electron beam is incident on a specimen, it broadens inside the specimen owing to the electron scattering. Such beam broadening can be estimated on the basis of the single scattering regime proposed by Goldstein et al. [8]. In this model, shown in Fig. 1.2, the incident electrons are assumed to be scattered once in the middle of the specimen of thickness t. Whereas beam broadening may be generally defined as a diameter of the base of the scattering cone at the end of the specimen, in the model proposed by Goldstein et al. beam broadening is defined as the diameter of the base of scattering cone (b) that contains 90% scattered electrons. It is given as

$$b = 6.25 \times 10^2 (\rho/A)^{1/2} (Z/E) t^{3/2} (\text{cm}) \tag{1.10}$$

where the units for b and t are centimeters and those of ρ, A, and E are grams per cubic centimeter, grams per mole, and kiloelectron volts (keV),

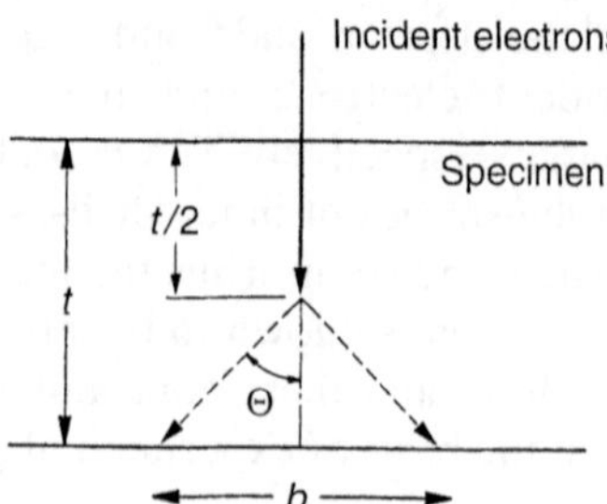

Fig. 1.2. Electron beam broadening in a specimen

respectively. According to this model, beam broadening is proportional to the specimen thickness t to the power of $\frac{3}{2}$, and it is proportional to the incident electron intensity but inversely proportional to the accelerating voltage. During analytical electron microscopy using a nanoprobe, it is important to keep this relation in mind. Beam broadening for specimens of crystalline Si and amorphous SiO_2 with various thicknesses has been measured and compared with estimations based on the above model [9].

1.1.2.4 Absorption Coefficient and Penetration Power

When an electron proceeds a small distance dt, the decrease in intensity dI is proportional to the intensity at the point and the distance dt, as given by

$$\mathrm{d}I = -\mu_0 I \mathrm{d}t \quad (1.11)$$

where μ_0, being a constant with the dimension of length in inverse, is called an absorption coefficent. The absorption coefficent divided by the density μ_0/ρ is the so-called mass absorption coefficient, which is specific for the material itself but does not depend on the assembling manner or the state of the material. The electron absorption process includes various inelastic scattering processes. The minus sign in the right-hand side of Eq. 1.11 indicates the decrease in intensity. By integrating Eq. 1.11, the electron intensity I for thickness t is given as:

$$I = I_0 e^{-\mu_0 t} \quad (1.12)$$

where I_0 is the incident electron intensity. The *penetration power*, which indicates the ability of electrons to move in the specimen, is given by the reciprocal of the absorption coefficient. Figure 1.3 indicates the relation between the accelerating voltage and the penetration power. In Fig. 1.3 the material is not specified, and the penetration power is normalized at 100 kV. It is seen that the penetration power basically increases with the increase in accelerating voltage.

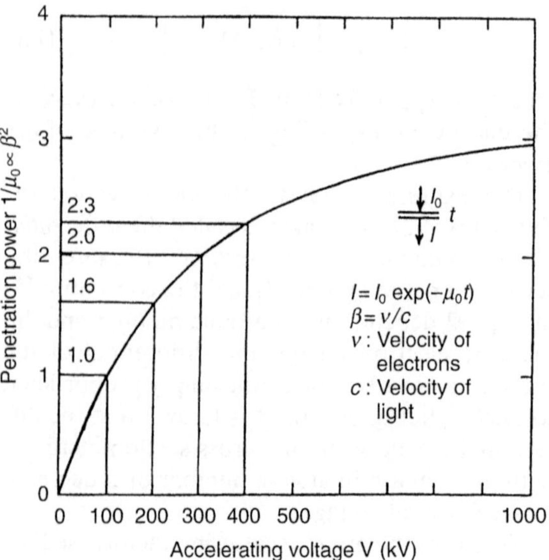

Fig. 1.3. Penetration power as a function of accelerating voltage

1.1.3 Simulation of Scattering Process

By using theoretical equations of the cross section and the mean free path of inelastic scattering, the scattering of the incident electrons and their trajectories in a specimen can be simulated. However, the scattering process and the mean free path are given as a statistical probability, so the specific behavior of electrons in the specimen can be simulated with random numbers. Figure 1.4 is an example of the simulation obtained by the software "Electron Flight Simulator," indicating the electron scattering processes for Al and Au thin foils of 3 μm thickness at the accelerating voltages of 1, 10, 100, and 400 kV. It is seen the range of electrons, which is the average distance the electrons travel until they stop, increase with the increase in accelerating voltage. This tendency is more prominent in the lighter element Al than in Au. In Fig. 1.5, detailed scattering processes of electrons in Al and Au with a thickness of 100 nm are shown for the accelerating voltage from 100–300 kV. At this thickness the electron beam tends to spread homogeneously with the increase in specimen thickness. It is also noted that broadening of the electron beam is suppressed more strongly in a light element and a high accelerating

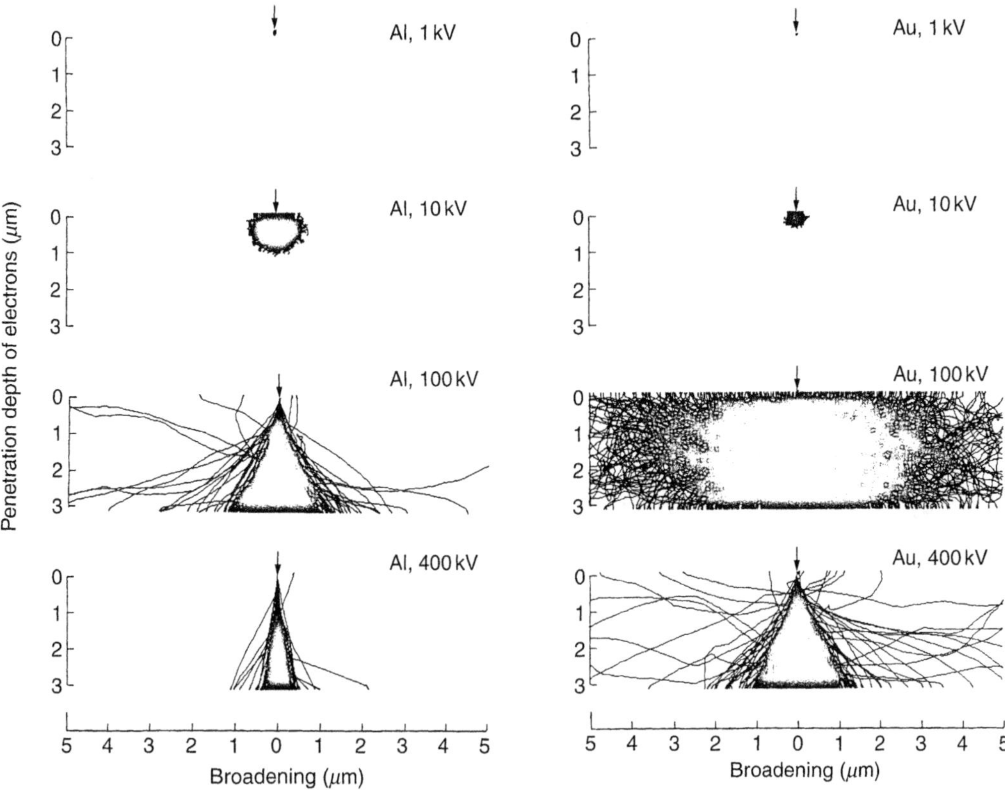

Fig. 1.4. Simulation of electron-scattering process for specimens (Al, Au) 3 μm thick at accelerating voltages of 1–400 kV

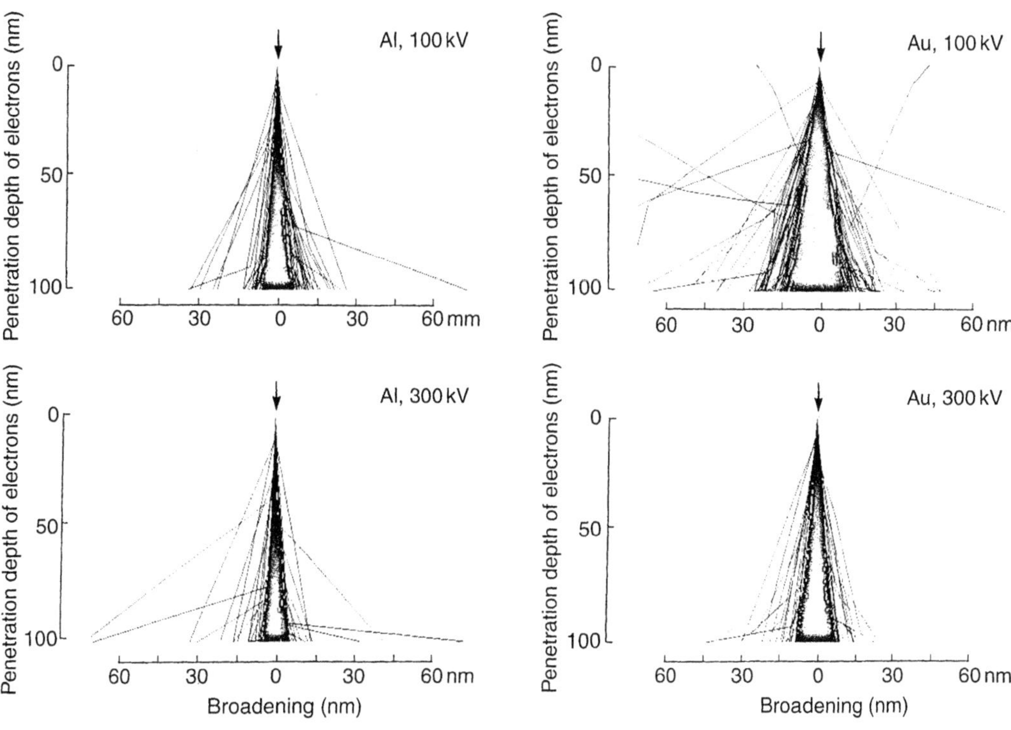

Fig. 1.5. Simulation of electron-scattering process for specimens (Al, Au) 100 μm thick at accelerating voltages of 100 and 300 kV

voltage, consistent with Eq. 1.10. Using the software, the X-ray intensity emitted through the inelastic scattering process is also calculated in the same line. Thus, simulation of electron scattering in the specimen is important for estimating the results of an electron scattering experiment prior to the experiment and interpreting the experimental data, taking into account electron beam broadening, X-ray emission, and so on.

1.2 Inelastic Electron Scattering and Analytical Electron Microscopy

1.2.1 Outline of Electron Energy-Loss Spectroscopy and Energy Dispersive X-ray Spectroscopy

The principles of EELS and EDS can be explained with one of the inelastic electron scattering processes (i.e., excitation of an inner-shell electron). Figure 1.6 shows the change in electronic structure due to excitation of an electron in the K shell. The resultant energy-loss spectrum and X-ray spectrum observed are shown at the bottom of Fig. 1.6. Here we consider the case where an incident electron gives energy to the specimen, and the electron in the K shell (1s orbital) is excited. Because the energy levels below the Fermi level are all occupied by electrons in the ground state, one of the electrons in the K shell can only transit to the unoccupied state above the Fermi energy. Thus, when the electron loses more energy than ΔE, which corresponds to the energy difference between the energy at the K shell and the Fermi energy, the probability of the transition from the K shell to the unoccupied density of states increases drastically, and eventually the sharp peak appears at energy ΔE in an energy-loss spectrum. In this excitation process of the inner-shell electron, the peak tends to accompany the tail in a higher energy region. From this shape, the peak appearing in the energy-loss spectrum is generally called an *edge.* Because the threshold energy of the edge is specific to each material, the specimen can be identified by its energy value ΔE. Also, information about the content of the element can be obtained from the integrated intensity of the edge. Furthermore, from an accurate value of the threshold energy and the shape of the edge, information about the chemical bond can be obtained. Interpretation and analysis of the edge are presented in detail in Chapter 3. There are also other excitation processes of the atom due to the incident electrons, such as the interband transition and collective excitation of valence electrons.

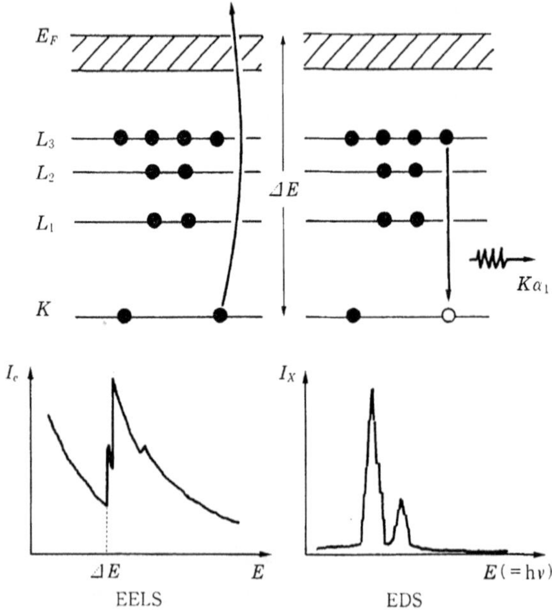

Fig. 1.6. Inner-shell electron excitation, a resultant electron energy-loss spectrum and an energy dispersive X-ray spectrum

When the atom changes from the excited state to the ground state, the surplus energy is emitted as a characteristic X-ray or an Auger electron. In both cases, one of the electrons in a higher energy level transits to the hole in the lower energy level in a manner satisfying the selection rule. Like EELS, the energy of the characteristic X-ray can be used to specify a constituent element, as the energy at the X-ray peak position is specific to each material. Also, the composition of the material can be evaluated from the integrated intensity. The characteristic X-ray emission resulting from transition of the electron from the L_3 shell to the K shell is illustrated in Fig. 1.6, and the resultant X-ray emitted is called a characteristic K_{α_1} X-ray. Several other characteristic X-rays are also frequently used for compositional analysis, such as K_{β_1} and L_{α_1}, corresponding to the transitions from the M_3 shell to the K shell and from the M_5 shell to the L_3 shell, respectively. Species of the transitions and their names are noted in Chapter 4.

As noted in the explanation above, EELS and EDS are related to the same excitation process (i.e., inner-shell electron excitation process), so it may be considered that similar information can be obtained from these two methods. However, owing to the large differences in the background height and the resolution of the spectra, the information obtained differs. For example, resolution of EELS is around 1 eV, whereas that of EDS is around 150 eV. To illustrate the situation, an electron energy-loss spectrum [10] and an X-ray spectrum obtained from a $YBa_2Cu_3O_y$ superconductor are presented in Figs. 1.7 and 1.8, respectively. Although the electron energy-loss spectrum can generally be observed in the energy range from 0 to about 2 keV in a conventional EELS system, only a limited energy region of 500–950 eV is presented in Fig. 1.7a. Small peaks of oxygen K-edge, an $Ba\text{-}M_{4,5}$ edge, and a $Cu\text{-}L_{2,3}$ edge are seen on a

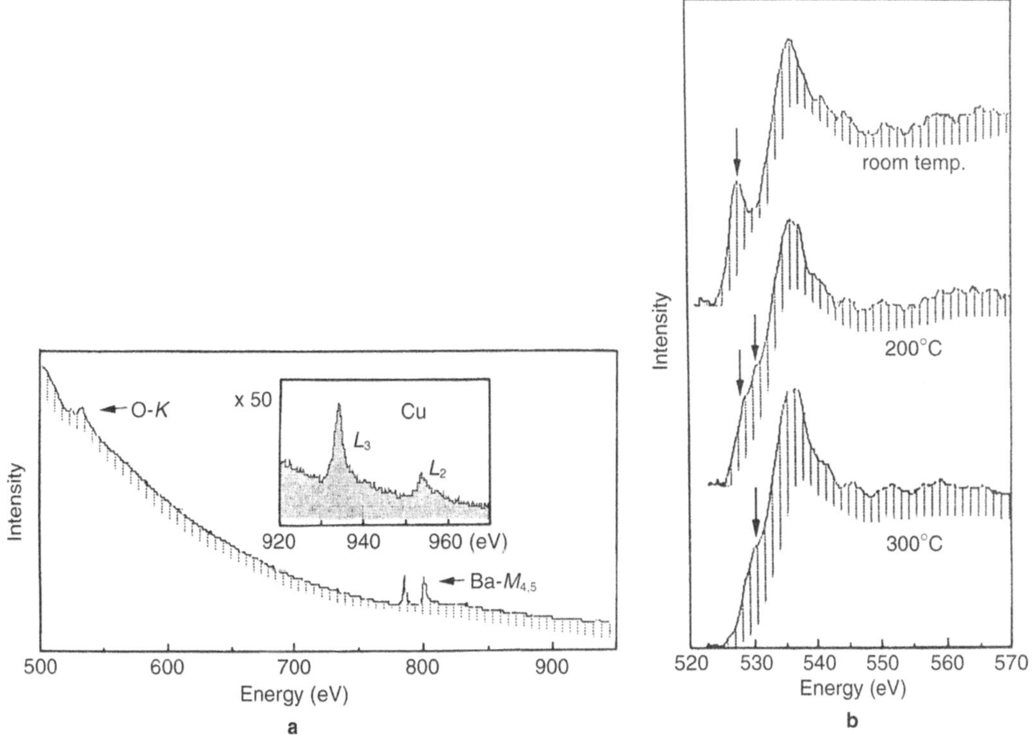

Fig. 1.7. Electron energy-loss specta of $YBa_2Cu_3O_y$. Energy ranges are 500–950 eV **a** and 520–570 eV **b**

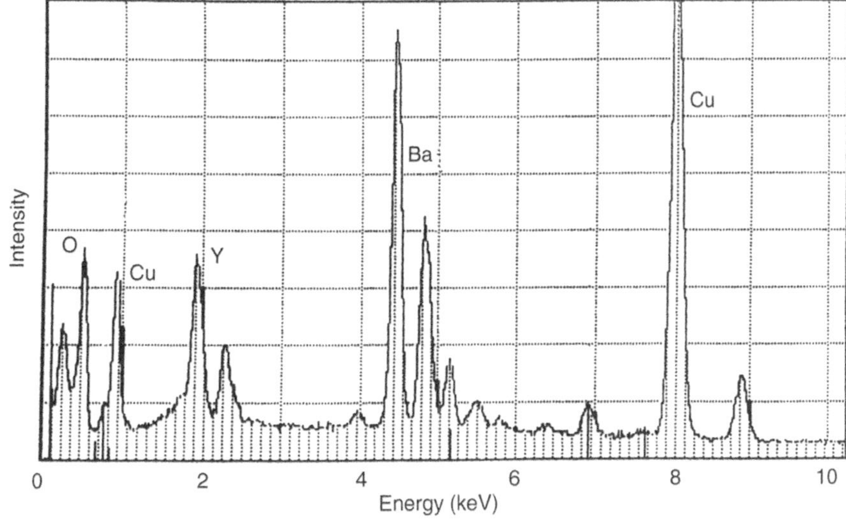

Fig. 1.8. Characteristic X-ray spectrum of $YBa_2Cu_3O_y$

large background. On the other hand, in Fig. 1.7b, three energy-loss spectra at the oxygen K-edge observed at different temperatures are presented in the energy range 520–570eV. Whereas the oxygen content decreases with the increase in temperature, resulting in decreased hole content, the peak at 528eV of the energy-loss spectrum decreases correspondingly with the decrease in hole content. Thus, it is seen that EELS shows not only the compositional information but also the information on the electronic state, especially unoccupied density of states (see Section 3.5.3.2). On the other hand, based on the X-ray spectrum in the energy range 0–10keV (a spectrum available up to 20keV), quantitative compositional information can be easily obtained owing to the low background of the spectrum.

1.2.2 Analytical Electron Microscopy and Materials Characterization

The analytical techniques presented in this book (e.g., EELS and EDS) and the information obtained by these techniques are shown in Table 1.1. The performance of analytical electron microscopes can be overviewed from Table 1.1. In contrast, in Table 1.2 the information necessary for characterizing materials is listed first, and the analytical techniques that may be used for such characterization are presented in the right-hand column. It is thus seen that the technique necessary for obtaining specific information about materials is not limited to one test, but several are available for gaining such information. For example, if information on composition is necessary, EDS, EELS, and Z-contrast imaging can be used. It should be noted, however, that the experimental conditions and accuracy are different from one another. Also, wtih a specific technique (e.g., EELS) the accuracy changes depending on the elements and composition. Thus, to choose an appropriate technique for the characterization, it is necessary to understand the principles of the technique (see the sections indicated in the parentheses in Table 1.2).

In practice, not a single technique but a combination of complementary techniques is generally useful for analyzing advanced materials. Electron diffraction patterns of Bi-based high-Tc super-

Table 1.1. Analytical methods and information obtained thereby.

Analytical methods	Information obtained
Electron energy-loss spectroscopy (EELS)—(see Chap. 3)	Electronic structure, composition, thickness
Energy dispersive X-ray spectroscopy (EDS)—(see Chap. 4)	Composition, impurity atom position, thickness
High-resolution electron microscopy (HREM) [1]	Atomic arrangement, lattice defect, surface morphology
Electron diffraction (ED)—(see Sect. 5.1)	Crystal structure, crystal orientation, thickness
Lorentz electron microscopy—(see Sect. 5.2)	Magnetic structure
Electron holography—(see Sect. 5.3)	Magnetic structure, thickness, inner potential
Scanning electron microscopy (SEM)—(see Sect. 5.4.1)	Surface morphology
Z-contrast method—(see Sect. 5.4.2)	Composition, atomic arrangement

Table 1.2. Information necessary and analytical methods to be utilized.

Information necessary	Analytical methods possible
Composition	EDS (see Sect. 4.3), EELS (see Sect. 3.3), HREM [1], Z-contrast (see Sect. 5.4.2)
Electronic structure	EELS (see Sect. 3.5.3.2)
Thickness	CBED (see Sect. 5.1.2), EELS (see Sect. 3.5.2), holography (see Sect. 5.3)
Crystal orientation	ED (see Sect. 5.1)
Surface morphology	SEM (see Sect. 5.4.1), HREM [1]
Atomic arrangement	HREM [1], ED (see Sect. 5.1), EDS (see Sect. 4.6)
Lattice defect	HREM, BF, DF, WB [1]
Magnetic structure	Lorentz EM (see Sect. 5.2), holography (see Sect. 5.3)

BF, bright-field method; DF, dark-field method; WB, weak-beam method; CBED, convergent beam electron diffraction

conductors and their energy dispersive X-ray spectra obtained in the same area as the electron diffraction patterns are presented in Fig. 1.9 [11]. In the electron diffraction patterns in Fig. 1.9 (A and B), the streaks along the c*-axis are seen, indicating the existence of irregularities in the stacking parallel to the c-plane. Comparing the characteristic X-ray peaks of these areas with those of C and D in Fig. 1.9, the Bi/Cu content ratio is small in the A and B areas. In other words, the Cu content increases with an increase in the stacking irregularities. To clarity the situation, a lattice image was observed in the area where strong streaks are observed in the electron diffraction patterns, as shown in Fig. 1.10. Here, the bright lines correspond to the Cu-O planes and the dark lines to the layers containing the heavy element Bi. The numbers in Fig. 1.10 indicate the number of successive Cu-O layers, and it is seen that successive stacking of three or four Cu-O layers occurs in some regions. Thus, in this specimen the composition changes from place to place owing to the irregular stacking of the Cu-O layer.

In the analysis noted above, the stacking manner in the local area could be analyzed by lattice imaging in addition to electron diffraction, and the composition could be evaluated by EDS. Thus, the combination of plural analytical techniques is important for structure characterization. With the recently developed analytical electron microscopes with a field emission gun, analysis can be carried out with a nanoprobe; and the compositional analysis in an area less than 1 nm and elemental mapping with the resolution better than 1 nm can now be carried out (see Fig. 4.8).

1.3 Hardware Controlled by Computer and Management of Analytical Data

To develop the performance of transmission electron microscopes, central processing units (CPUs) are introduced into transmission electron microscopes extensively these days. For example, the data of the electron current in various lens systems (see Section 2.1) are recorded in the CPU and are easily changed or reset to the optimum conditions or the initial conditions. Also, by using the CPU, functions such as autofocusing and autostigmatic adjusting can be used. Furthermore, one can utilize the minimum dose system (MDS) by which one can specify a region, focus it, and

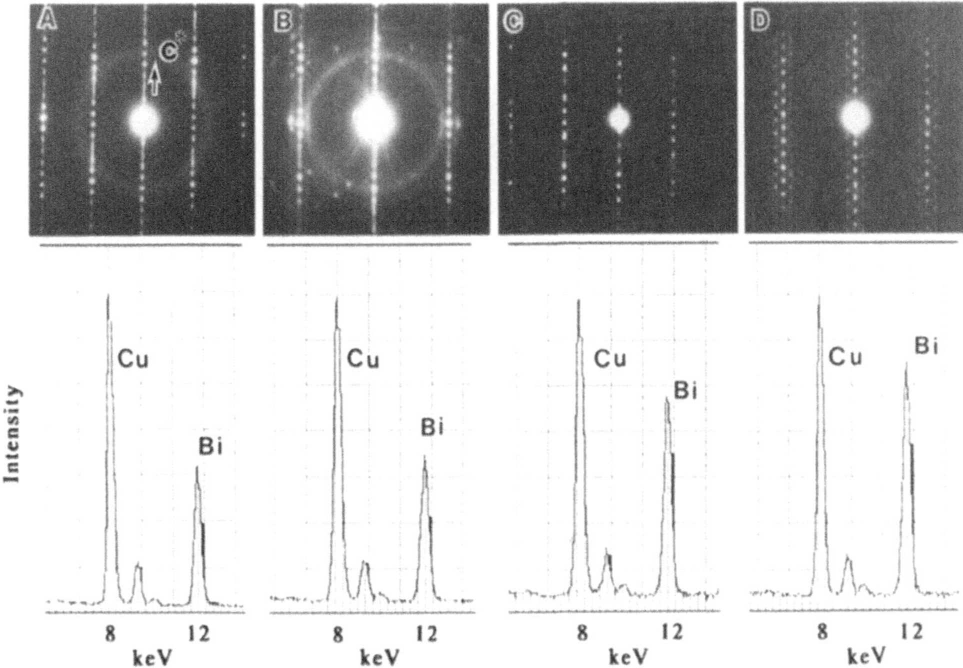

Fig. 1.9. Electron diffraction patterns and characteristic X-ray spectra of a Bi-based high-Tc superconductor

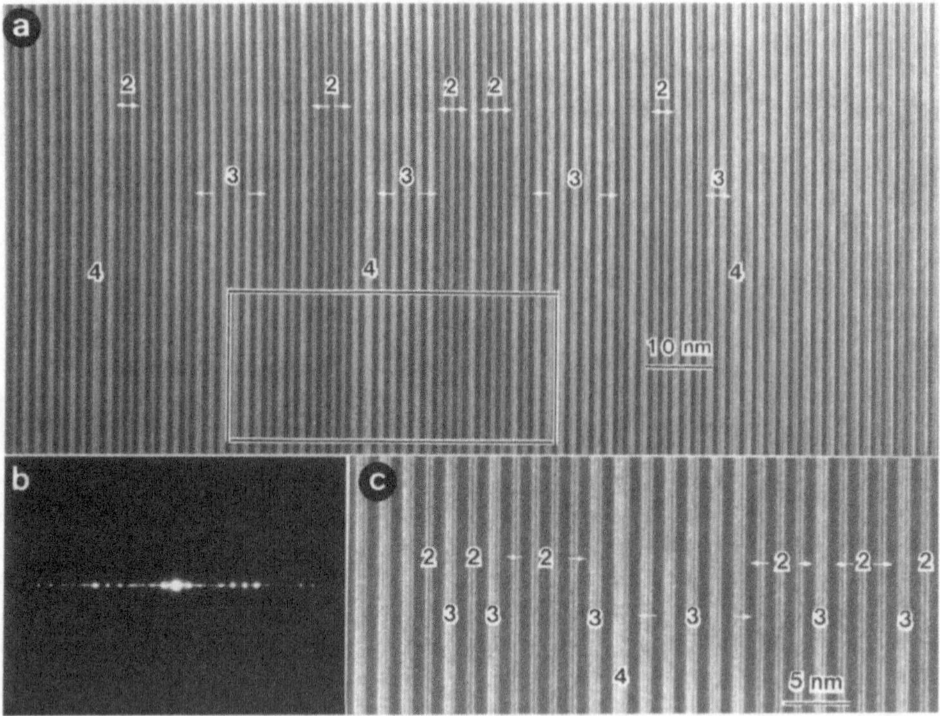

Fig. 1.10. **a** Lattice image and **b** electron diffraction pattern of a Bi-based high-Tc superconductor. **c** Enlarged image of the framed region in **a**

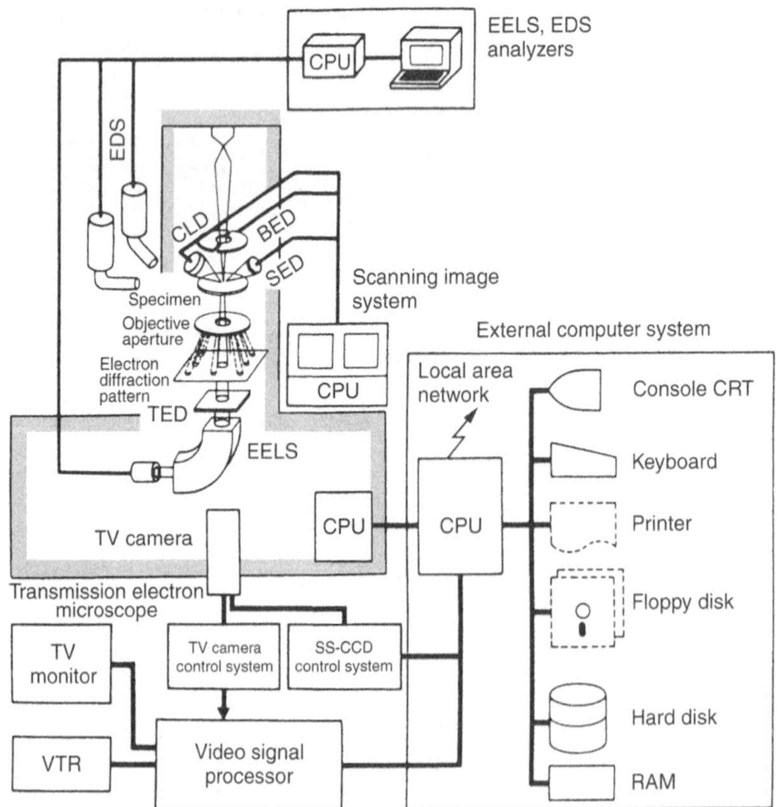

Fig. 1.11. Analytical electron microscope and its peripheral instruments. *CRT*, cathode ray tube; *SS-CCD*, slow-scan charge-coupled device

take a picture systematically with minimum electron intensity, minimizing the electron irradiation damage.

Figure 1.11 diagrams a transmission electron microscope and peripheral analytical instruments. As can be seen, the CPUs are used not only in the transmission electron microscope itself but they are also used to control the peripheral analytical instruments and manage analytical data. For example, in addition to the EELS and EDS systems, detector systems for secondary electrons, back-scattered electrons, and light quanta emitted by cathode luminescence are controlled by a personal computer (PC) and an engineering work station (EWS). In general, the data obtained are stored at once in random-access memory (RAM) and finally are recorded on a compact disk (CD) or magnetooptical disk (MO). Some of the data are output through the printer. Because electron microscopic images and electron diffraction patterns are obtained as digital data through a slow-scan CCD camera and an imaging plate system, analytical microscope data are analyzed with these images and diffraction patterns through a PC, an EWS, or both and are transferred to other computers through the network system.

For analyzing the digital data obtained, on-line and off-line software is available these days. Furthermore, various databases of electron microscope images and electron microscopy have been accumulated utilizing the internet [12, 13].

References

1. Shindo D, Hiraga K (1998) High-resolution electron microscopy for materials science. Springer, Tokyo
2. Wentzel G (1927) Zwei Bemerkunger über die Zerstreuung Korpuskularer Strahlen als Beugungserscheinung. Z Physik 40:590
3. Mott NF, Massey HWW (1965) The theory of atomic collision. Oxford University Press, Oxford
4. Cosslett VE, Thomas RN (1964) Multiple scattering of 5–30 keV electrons in evaporated metal films. I. Total transmission and angular distribution. Br J Appl Phys 15:883
5. McKinley WA, Freshbach H (1948) The Coulomb scattering of relativistic electrons by nuclei. Phys Rev 74:1759
6. Powell CJ (1976) Cross sections for ionization of inner-shell electrons by electrons. Rev Mod Phys 48:33
7. Powell CJ (1976) Use of Monte Carlo calculations. National Bureau of Standards Special Publication 460. NBS, Washington, DC, p 97
8. Goldstein JI, Costley JL, Lorimer GW, Reed SJB (1977) Quantitative X-ray analysis in the electron microscope. In: Johari O (ed) Proceedings of the workshop on analytical electron microscopy, scanning electron microscopy, Chicago, vol 1, p 315
9. Lee C-W, Kidu S, Oikawa T, Shindo D (2001) Estimation of electron beam broadening in specimen for analytical electron microscopy. In: Proceedings Microscopy and Microanalysis, Long Beach, California, vol 7. Springer, New York, p 204
10. Shindo D, Hiraga K, Hirabayashi M, Kikuchi M, Syono Y, Furuno S, Hojou K, Soga T, Otsu H (1989) In situ observation of oxygen K-edge fine structure of $YBa_2Cu_3O_{7-y}$ by EELS. J Electron Microsc 38:155
11. Shindo D, Hiraga K, Hirabayashi M, Kobayashi N, Kikuchi M, Kusaba K, Syono Y, Muto Y (1988) Analytical electron microscopic study of high-T_c superconductor Bi-Ca-Sr-Cu-O. Jpn J Appl Phys 27:L2048
12. Taniyama A, Shindo D, Hiraga K, Oikawa T, Kersker M (1997) Database of electron microscope images on the World Wide Web. In: Proceedings Microscopy and Microanalysis, Cleveland. Springer, New York, p 1105
13. Shindo D, Ikematsu Y, Lim S-H, Yonenaga I (2000) Digital electron microscopy on advanced materials. Mater Characterization 44:375

2. Constitution and Basic Operation of Analytical Electron Microscopes

The constitution, functions, and principles of an analytical electron microscope are explained in this chapter. For observing electron microscope images and diffraction patterns and for carrying out various extensive analyses, it is important to set an analytical electron microscope to be in optimal operating condition by learning the principles of its constituent units and its appropriate operating method. Here, the basic configuration of a transmission electron microscope is noted first, and its constituent units are described in turn. Alignment of various lens axes and adjustment of their astigmatism are explained based on these explanations.

2.1 Basic Constitution of Analytical Electron Microscopes

Figure 2.1 shows the appearance of a conventional analytical electron microscope. Figure 2.2 shows a cross section of the electron microscope together with the names of its constituent parts.

Electrons are emitted from an electron gun installed at the top of the column in the transmission electron microscope. The column is kept at high vacuum by evacuating the air. The electrons emitted are accelerated in the acceleration tube and then pass through illumination lenses and are incident on a specimen. After passing through the specimen, the electrons form an image by appropriate action of the objective lens. The enlarged image is formed via image-forming lenses. The final image formed on the fluorescent screen is observed through a window of a viewing chamber, and the image is recorded on photo-film in the camera chamber.

Following the path of an electron beam in the microscope column, the electron microscope is divided as follows.

1. Electron gun (electron source)
2. High voltage generator and an acceleration tube
3. Illumination lens system and a deflector
4. Specimen holder and a stage
5. Image-forming lens system
6. Viewing chamber and camera chamber

These units are successively described in detail below.

An electron energy-loss spectrometer (EELS) and an *energy dispersive X-ray spectrometer* (EDS) are shown with the electron microscope in Fig. 2.1 and are described in detail in Chapters 3 and 4, respectively. The function and utility of various vacuum pumps in the electron microscope is presented in Appendix C.

2.1.1 Electron Gun

An *electron gun* generates electrons and is located at the top of the microscope column (Fig. 2.2). Characteristics of the electron beam (e.g., the diameter of a electron beam, the width of energy spread) depend on the type of electron gun. There are two emission types of electron gun: the thermionic *emission* type and the *field emission* type. The tungsten (W) hairpin filament of a thermionic emission type electron gun was widely used in conventional transmission electron microscopes for a long time. The lanthanum hexaboride (LaB_6) single-crystal filament of a thermionic emission type electron gun, which is the same as a hairpin filament but with higher brightness, has been widely used recently. The *field emission gun* (FEG), which can generate the electrons in higher brightness and higher coherence, has been widely introduced into analytical electron microscopes. There are two types of FEG. One is a *cold FEG*, and the other is a *thermal FEG*. Table 2.1 shows characteristics of these electron guns.

2.1.1.1 Thermionic Emission Gun

Figure 2.3 shows a thermionic emission electron gun and its assembly. The filament, which

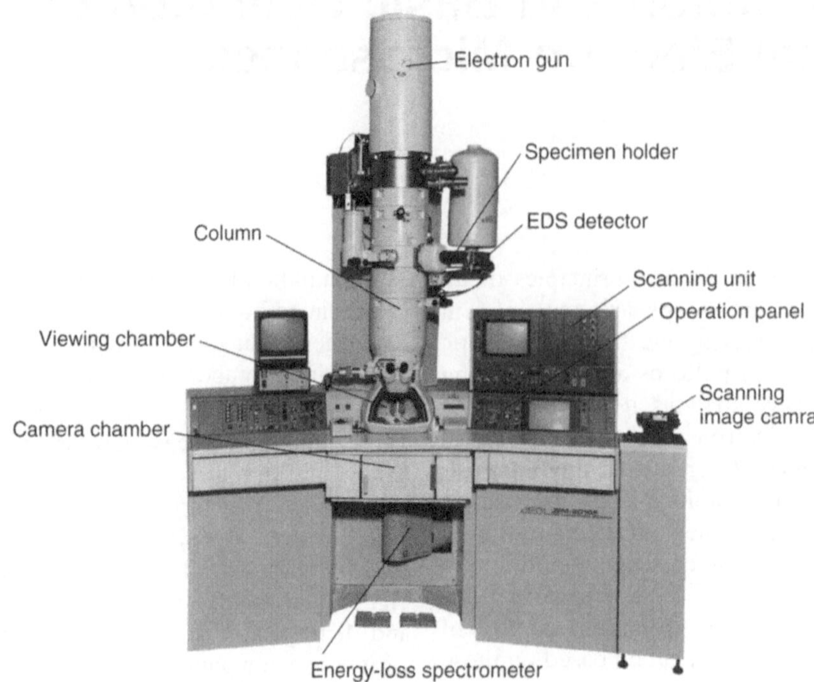

Fig. 2.1. Appearance of a transmission electron microscope (TEM) (JEM-2010F). EDS, EELS, scanning unit and TV-camera are installed

Table 2.1. Comparison of characteristics of various electron guns.

	Thermionic emission		Field emission		
			Thermal FEG		Cold FEG
Characteristic	W	LaB_6	ZrO/W (100)	W (100)	W (310)
Brightness ($A/cm^2 sr$) at 200 kV	~5×10^5	~5×10^6	~5×10^8	~5×10^8	~5×10^8
Source size	50 μm	10 μm	0.1–1.0 μm	10–100 nm	10–100 nm
Energy spread (eV)	2.3	1.5	0.6–0.8	0.6–0.8	0.3–0.5
Conditions for usage					
Pressure (Pa)	10^{-3}	10^{-5}	10^{-7}	10^{-7}	10^{-8}
Temperature (K)	2800	1800	1800	1600	300
Emission					
Current (μA)	~100	~20	~100	20–100	20–100
Stability for short time	1%	1%	1%	7%	5%
Stability for long time	1%/h	3%/h	1%/h	6%/h	5%/15 min
Current efficiency	100%	100%	10%	10%	1%
Maintenance	No need	No need	Take some time for setup	Build up several times for a new tip	Flashing necessary every few hours
Price/operation	Cheap/ simple	Cheap/ simple	Expensive/ easy	Expensive/ easy	Expensive/ complicated

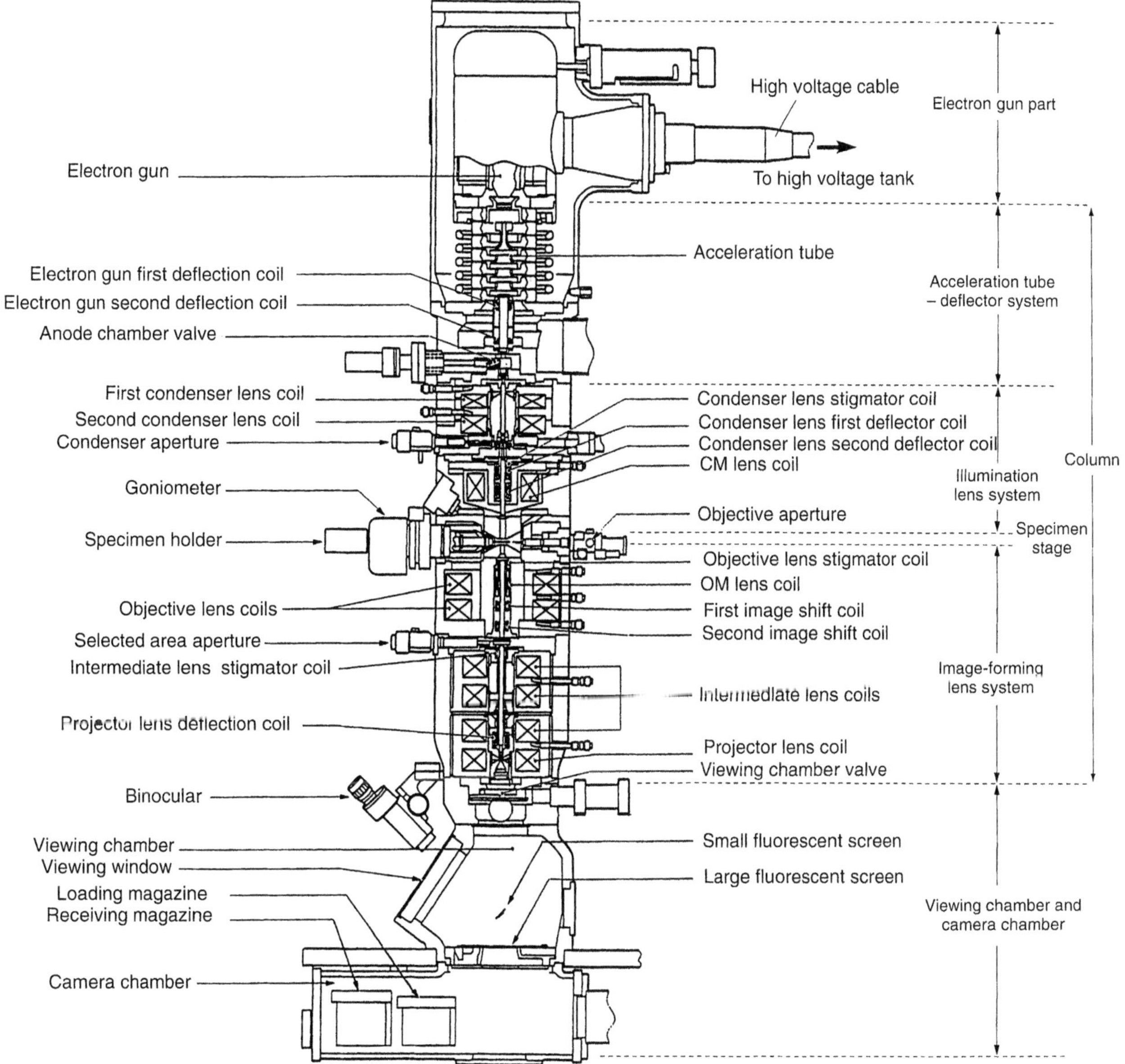

Fig. 2.2. Cross section of a column in a TEM (JEM-2010F). *CM*, condenser minilens; *OM*, objective minilens

generates electrons, is made of a tungsten (W) wire or an LaB_6 single crystal. As shown in Table 2.1, the LaB_6 filament should be used under a higher vacuum than the tungsten filament; the LaB_6 filament provides higher brightness, smaller electron source, and small amount of energy spread. Thus, the LaB_6 filament is more suitable than the tungsten hairpin filament for analytical electron microscopy. Figure 2.4 shows an electrical circuit of the electron gun of the thermionic emission type. Electrons generated from the filament (cathode) are accelerated in an acceleration tube (anode). The anode is set at ground voltage, and negative high voltage is applied to the cathode. The so-called *bias voltage* (lower than in the cathode) is applied to the Wehnelt electrode, which is located below the filament. The bias voltage controls the emission current and electron trajectories around an electron crossover point. The system supplying the bias voltage shown in Fig. 2.4 is called a *self-bias system*. A current of the same amount as the emitted electrons flows into bias resistance. The voltage generated by the bias resistance is supplied between the filament and

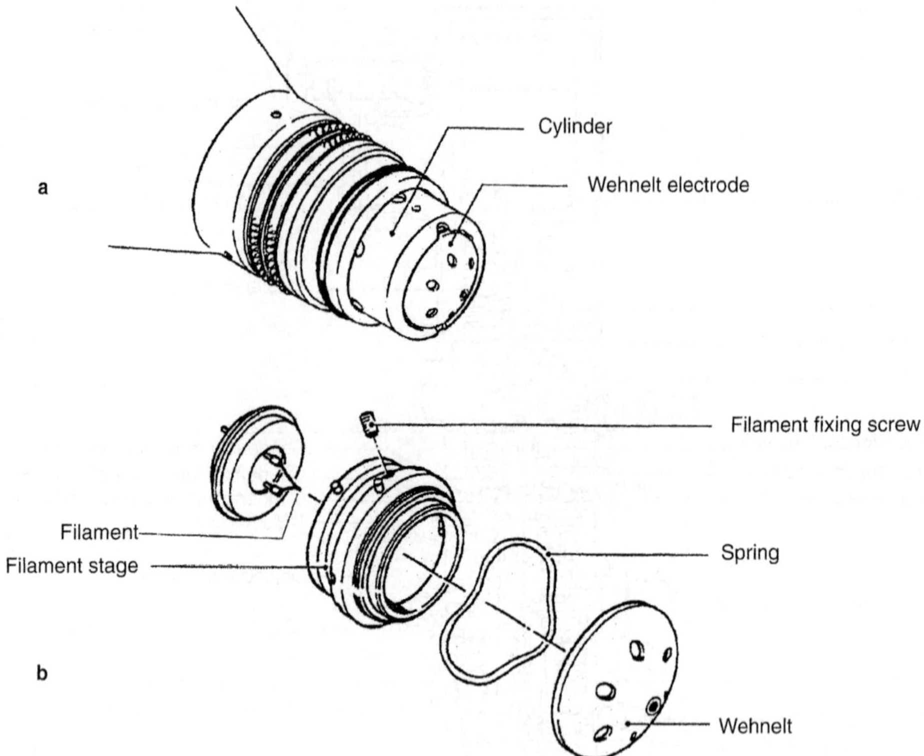

Fig. 2.3. Appearance **a** and structure **b** of a thermionic-type electron gun

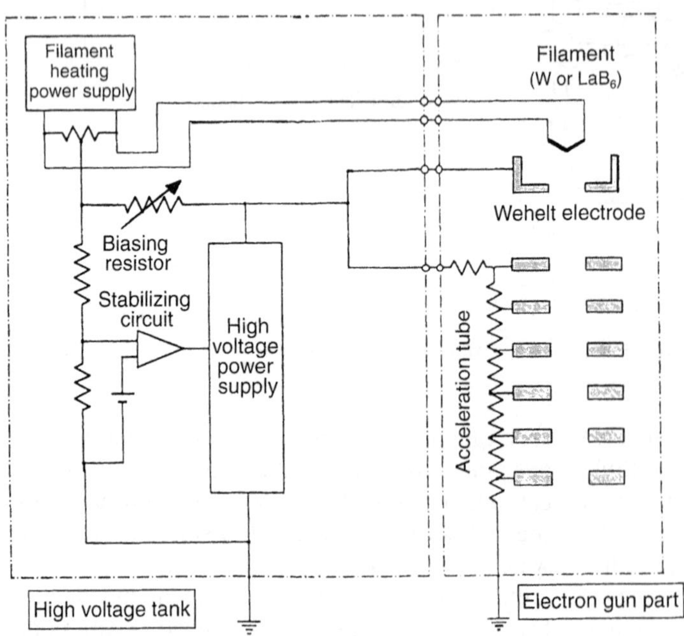

Fig. 2.4. Electric circuit of electron gun of thermionic type

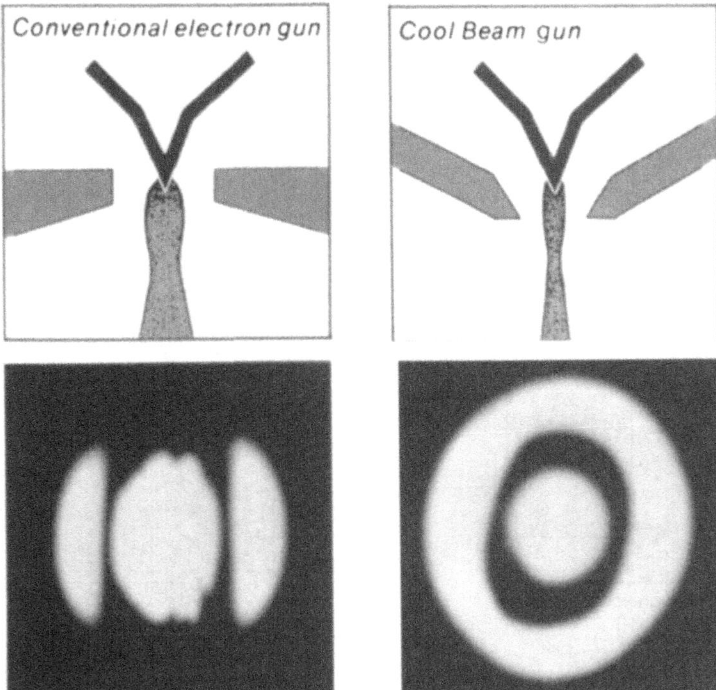

Fig. 2.5. Electron current (*top*) and filament patterns (*bottom*) of a tungsten hairpin filament

the Wehnelt electrode. This is a kind of feedback system for the emission current. Thus, the emission current is controlled automatically and stabilized. The method for adjusting the electron gun is described in Section 2.2.1.1. On the other hand, the *fixed bias system* supplies the bias voltage without bias resistance. The emission current is controlled directly, being independent of the emission current. The system utilizing the advantages of both self-bias and fixed bias methods is called a *half-fixed bias system*. Figure 2.5 shows the distribution of electrons emitted from the electron gun and emission patterns (filament patterns projected on a fluorescent screen) of a tungsten hairpin filament. The electrons emitted from a heated filament tip are condensed by the Wehnelt electrode. The shape of the emission patterns depends on the shape of the Wehnelt electrode and the condition of the bias voltage.

2.1.1.2 Field Emission Gun

Electrons in metals pass through a potential barrier by a tunneling effect and can be emitted from the surface of metals because the potential barrier of the metal–vacuum boundary thins when a strong electrical field is applied to the surface of the metal. This phenomenon is called *field emission*. The cathode fabricated with a sharply pointed shape with a 0.1 μm radius of curvature to localize the electric field is called an emitter or a tip. The FEG produces about 100 times higher electron brightness than that of the thermionic emission gun made with an LaB_6 single crystal filament and provides an extremely small electron source. Owing to the characteristics mentioned above, it is easy to make a small probe and derive high brightness; hence, the FEG is now utilized widely in analytical electron microscopy. Moreover, the high coherence of electrons produced with the FEG makes electron holography study possible (see Section 5.3).

Cold FEG. A cold FEG employs tungsten with the surface of the (310) plane as an emitter. The emitter works at room temperature without heating. Because the energy spread is as small as 0.3–0.5 eV, it is expected to obtain high-energy resolution in EELS. On the other hand, contamination of the residual gas appears on the surface of the emitter. It generates emission noise or instability of the emission (or both). The emission current decreases slowly with an increase in the contamination layer. Thus, regular maintenance, the so-called flashing procedure, is necessary.

Thermal FEG. When heating the emitter to a lower than thermal electron emission temperature of 1600–1800 K under a strong electric field, elec-

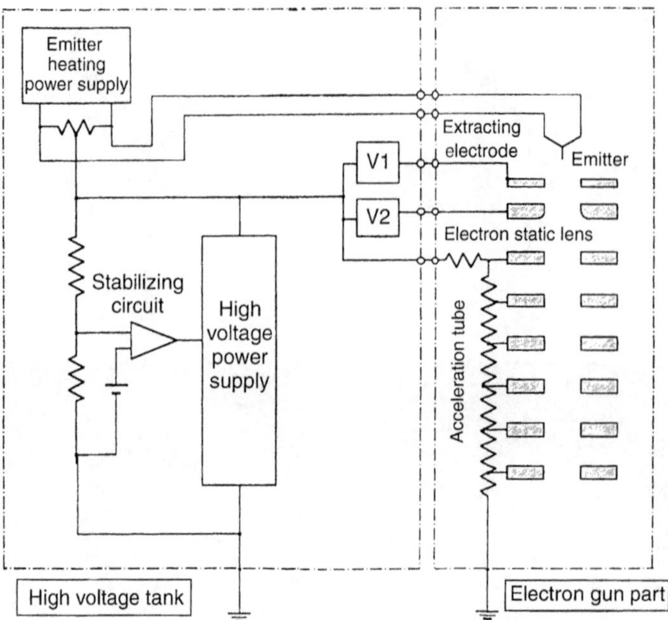

Fig. 2.6. Electrical circuit of thermal field emission gun (FEG)

trons are emitted that pass through a potential barrier that decreases because of the field applied. This phenomenon is called the *Schottky effect.* It has the disadvantage of a large energy spread (0.6–0.8 eV) because of heating the emitter compared to that of a cold FEG. On the other hand, it has less emission noise and provides a stable emission current without flashing because there is no adsorption of contamination on the emitter. Figure 2.6 shows an electrical circuit of the thermal FEG and its power supply. An extraction electrode and an electrostatic lens are used in the thermal FEG instead of the Wehnelt electrode. Figure 2.7 shows an example of beam intensity distribution with the full width at half maximum (FWHM) of 0.5 nm obtained by a thermal FEG.

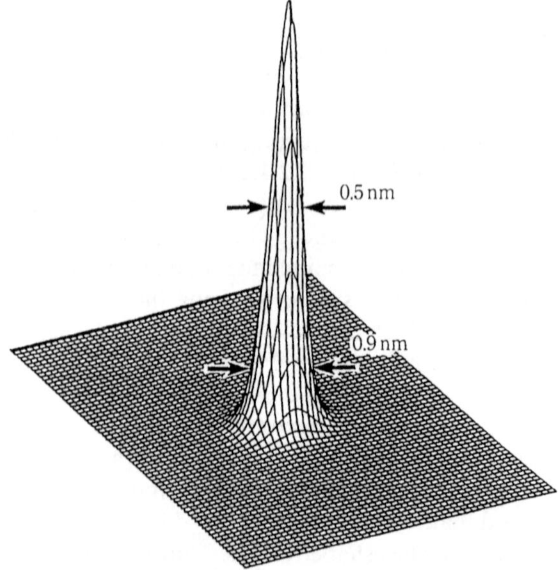

Fig. 2.7. Intensity profile of an electron beam obtained with a thermal FEG. The profile was observed with the imaging plate in a JEM-2010F TEM at accelerating voltage of 200 kV

2.1.2 High Voltage Generator and Acceleration Tube

A device generating a high voltage that is utilized to accelerate electrons in the electron gun is called a *high voltage generator* or a *high tension generator*. Housing for the high voltage generator is called a *high voltage tank* or *high tension tank*. On the other hand, a unit to accelerate electrons with the use of high voltage is an *acceleration tube*. The high voltage generator and acceleration tube are connected with a high voltage cable, as shown in Fig. 2.2. A variation of the voltage generated in a high voltage generator should be kept as small as possible because it causes defocusing of electron microscope images due to the chromatic aberration (see Section 2.1.5.1). *Cockcroft-Walton's high voltage circuit* (CWC) is employed as a high voltage generator in transmission electron microscopes. The principle of the circuit is described below.

As shown in Fig. 2.8, the CWC is composed of a transformer T, condensers $C_{A1}, C_{A2}, C_{A3}, \ldots, C_{B1}, C_{B2}, C_{B3}, \ldots$, and diodes $D_1, D_2, D_3, \ldots$. The condensers and diodes are connected, forming the step-like structure. When alternating current (AC) voltage is applied at the input terminal of the transformer T, a peak voltage generated at the output terminal of T is defined as V. When the voltage at A_0 becomes lower than that at B_0, the electric current flows to A_1 through diode D_1. At this moment the condenser C_{A1} is charged up with electrons $Q = CV$ (where C is the capacity of C_{A1}). At the next moment, the voltage at A_0 becomes $+V$ when the AC phase is shifted. Then the voltage at A_1 is increased up to 2 V. By repeating this process, voltages at B_1, B_2, and B_3 become 2 V, 4 V, and 6 V, respectively. In practice, the condensers are not charged fully, so the maximum voltage obtained V_{max} for the n-stage circuit is given as follows:

$$V_{max} = 2nV - \frac{2}{3} n^3 \frac{1}{fC} \tag{2.1}$$

where f is a frequency of AC.

Figure 2.9 shows a high voltage generator of 200 kV. The high voltage generated with this device is supplied to the acceleration tube through the high voltage cable. The acceleration tube

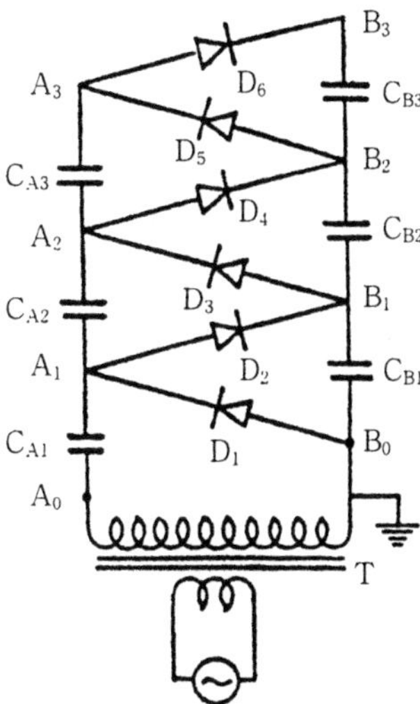

Fig. 2.8. Cockcroft-Walton's high voltage circuit (CWC)

accelerates electrons using this voltage. A *multi-stage acceleration electrode* is employed in transmission electron microscopes with an accelerating voltage higher than 200 kV. In the case of a high voltage electron microscope (HVEM) (higher than several hundred kilovolts), more than 15 stages of acceleration electrodes are used.

2.1.3 Illumination Lens System and a Deflector

The illumination lens system and the deflector provide functions by which the electrons converge on a specimen. The analytical electron microscope has the advantage of carrying out analyses in a small area (<1 nm in diameter). The advantage is achieved by the function of the illumination lens system, which can produce a small-probe and change the illuminating condition from parallel to convergent. Figure 2.10 shows a convergent illumination lens system. Figure 2.10a shows the condition of the parallel illumination on a wide area of a specimen, providing highly coherent electron illumination. For this condition, the *condenser minilens* (CM lens) is strongly excited, and electrons are focused on the prefocal point of the objective prefield. Figure 2.10b shows the conditions of small-probe illumination. The CM lens is turned off, and electrons are focused on the specimen by the objective prefield. The illumination angle (α_1) becomes large, providing a higher intensity of electrons. This condition is suitable for analyzing a small area. Figure 2.10c shows the conditions for a smaller illumination angle (α_2) utilizing a small condensor aperture. Under this condition, a small-diameter probe with relatively high coherence in illumination is used. For the conditions shown in Fig. 2.10b and Fig. 2.10c, it is possible to change the illumination angle (α), keeping the probe diameter constant, by changing the excitation of the condenser lenses and the CM lens. This is suitable for observing *convergent beam electron diffraction* (CBED) patterns.

A deflector for electron deflection is used for beam alignment, beam tilting, beam shifting, beam scanning, and so on. The deflector, composed of a pair of *deflection coils*, is called the *double-deflection* system and provides ease of operation. The principle of the double-deflection system is as follows: It works with a pair of deflection coils, as shown in Fig. 2.11. For making the beam tilt of θ_2 against the specimen, the electron

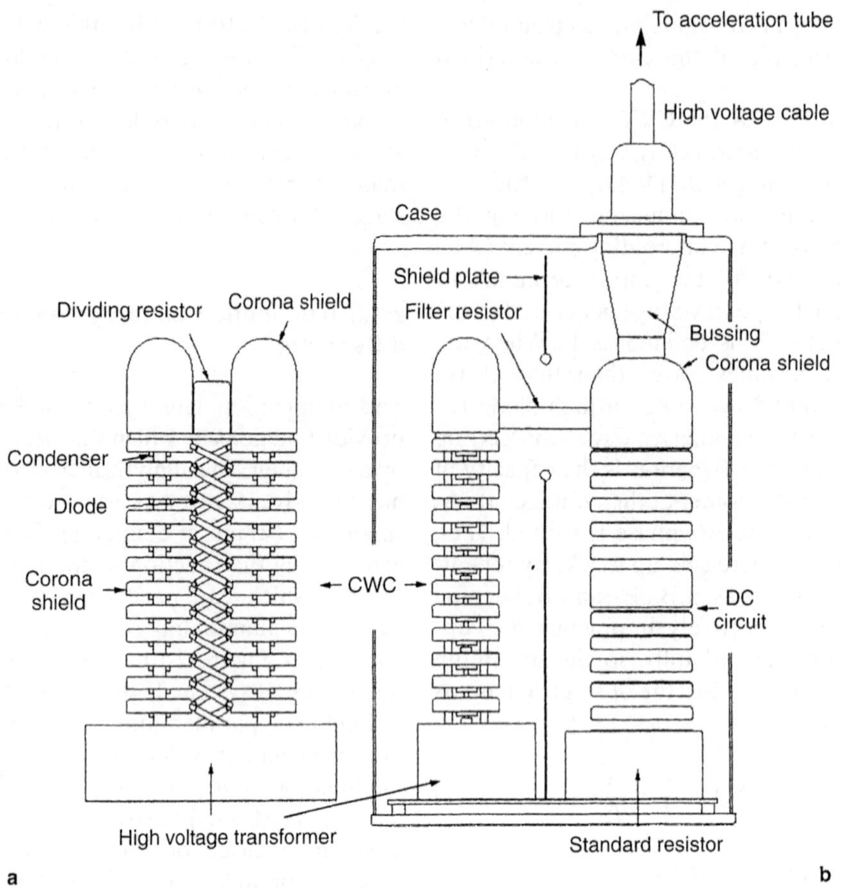

Fig. 2.9. Configuration of a high voltage generator (200 kV). **a** Cross section of Cockcroft-Walton's high voltage circuit (CWC). **b** High voltage tank containing CWC and DC circuit

beam is tilted by θ_1 first for the reverse direction by the first-stage deflection coil (DEF1) and tilted back by the second-stage coil (DEF2). There is a geometrical relation between θ_1 and θ_2 as given by

$$\tan\theta_2 = \frac{l_1}{l_2}\tan\theta_1 \tag{2.2}$$

where l_1 is the distance between DEF1 and DEF2; and l_2 is the distance between DEF2 and a specimen. By presetting the current ratio for DEF1 and DEF2 with the balance adjuster, the beam-tilt angle θ_2 can be changed by only one volume controller. In this system, the beam position on the specimen is kept the same, whereas the tilting angle can be changed. Similarly, the tilting angle can be kept constant and the beam position changed, as shown in Fig. 2.12. Thus, adjusting the deflector balance allows us to control the beam tilt (or the beam shift) with only one knob, independent of the beam shift operation (or the beam tilt operation). The deflectors are used not only in the illumination system but also in the electron gun, the image-forming system, and the projection system for beam alignment.

2.1.4 Specimen Holders

A specimen-supporting unit to be inserted into a transmission electron microscope is called a *specimen holder*, where a specimen of 3 mm diameter is fixed with a specimen fixer. For high-resolution electron microscopy and electron diffraction studies, it is necessary to align the specimen orientation accurately. Thus, a double-tilt specimen holder with a mechanism that allows the tilts for two axes to be perpendicular to each other is utilized for microstructure analysis of crystalline specimens.

There are two types of specimen holder. One is a *top-entry design* and the other is a *side-entry design*. The top-entry type has the mechanics for

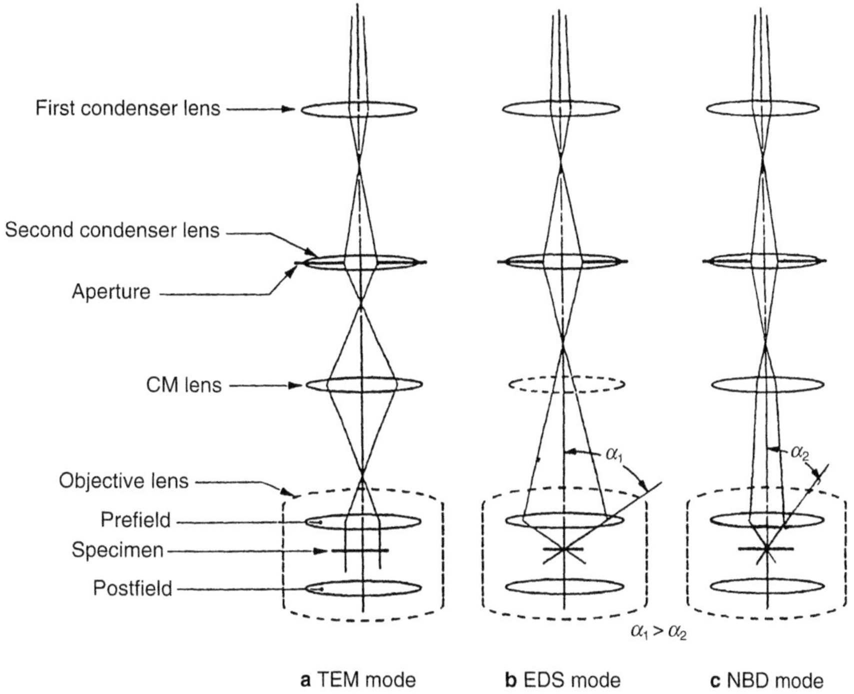

Fig. 2.10. Illumination lens systems. **a** TEM mode. **b** Energy dispersive X-ray spectroscopy (*EDS*) mode. **c** Nano-beam diffraction (*NBD*) mode

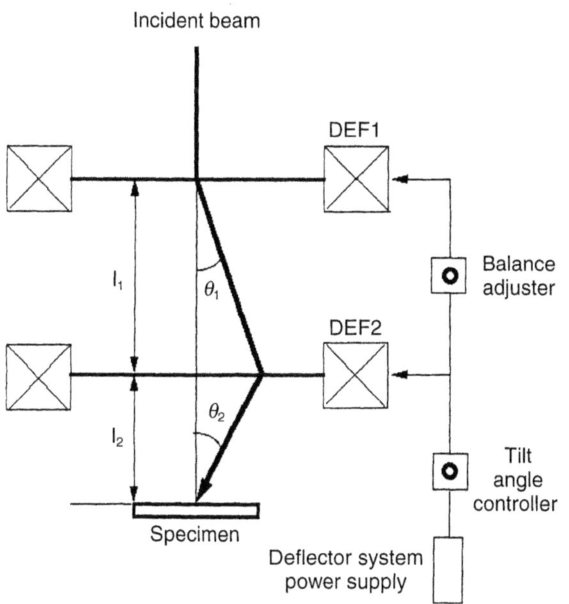

Fig. 2.11. Principle of a double-deflection system for beam tilt. *DEF1*, first-stage deflection coil; *DEF2*, second-stage deflection coil

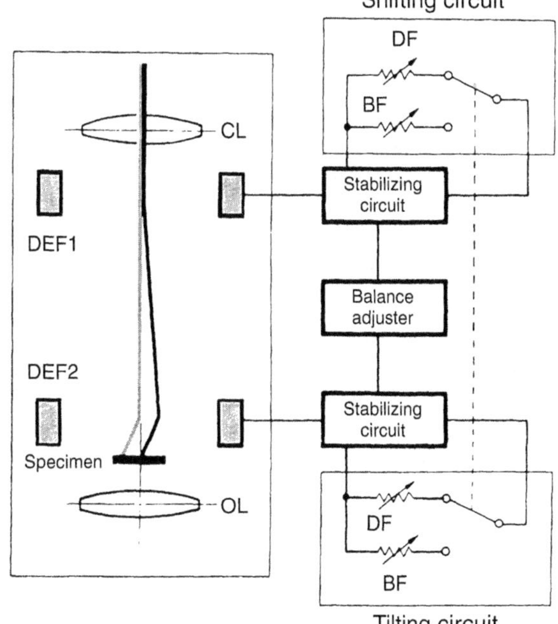

Fig. 2.12. Principle of a double-deflection system for beam shift. *DF*, dark-field imaging; *BF*, bright-field imaging

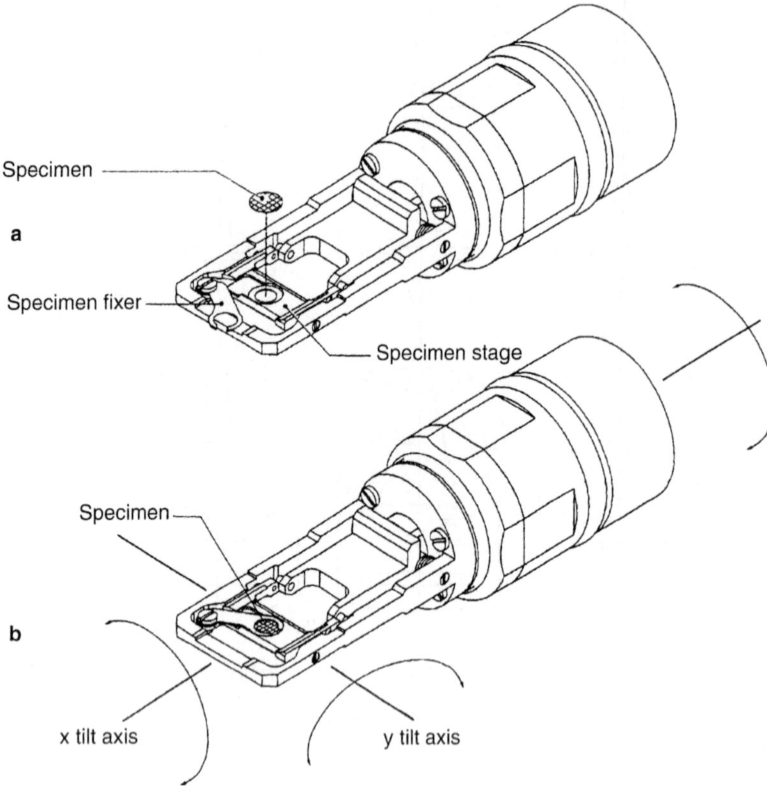

Fig. 2.13. Structure and mechanics of double tilting in a specimen holder of a side-entry type

inserting a specimen from the upper side of the polepiece, and the side-entry type has the mechanics for inserting it from the side of the polepiece. The top-entry type has the advantage of being stable to external vibration and thermal variation, as the specimen holder is dropped into a stable setting in the microscope column, and its structure is highly symmetrical. The top-entry type has been widely used for high-resolution observations. However, it has the disadvantage of a small tilting angle and the difficulty of introducing analytical capabilities. For these reasons a side entry-type specimen holder has been employed in analytical electron microscopes recently. This holder has the advantage of high sensitivity to signals such as X-rays and back-scattered electrons from the upper side of the specimen; it also allows large tilting angles of the specimen. Figure 2.13 shows the structure and mechanics of double tilts in a specimen holder for a side-entry type.

For EDS analysis, the analytical specimen holder composed of beryllium is used to minimize the hard X-ray background (a high-energy blemsstrahlung X-ray generated from a specimen by electron irradiation). Note, however, that beryllium is a poison, so touching a beryllium tip with the hand is prohibited.

There are various specimen holders (Fig. 2.14), such as a heating holder, a cooling holder to observe structural changes due to phase transformation with temperature, a mechanical testing holder for in situ observation of dislocation by adding tensile stress to a specimen, and so on. The heating holder has a molybdenum tip, which allows observation of a specimen at a temperature of about 800°C. There are two types of cooling holder. One uses liquid nitrogen (boiling point –195.8°C), so it is possible to observe a specimen at temperatures down to about –180°C. The other uses liquid helium (boiling point 4.21 K, –268.94°C), and specimens can be observed at temperatures down to about –250°C.

2.1.5 Image-Forming Lens System

2.1.5.1 Objective Lens

The *objective lens*, the first-stage lens, forms an image when electrons pass through the specimen. The image quality of the transmission electron microscope is determined mainly by the performance of the objective lens. It is composed of *lens coils*, a *magnetic circuit* (*yoke*), and a *polepiece*. The shape of the polepiece determines the optical

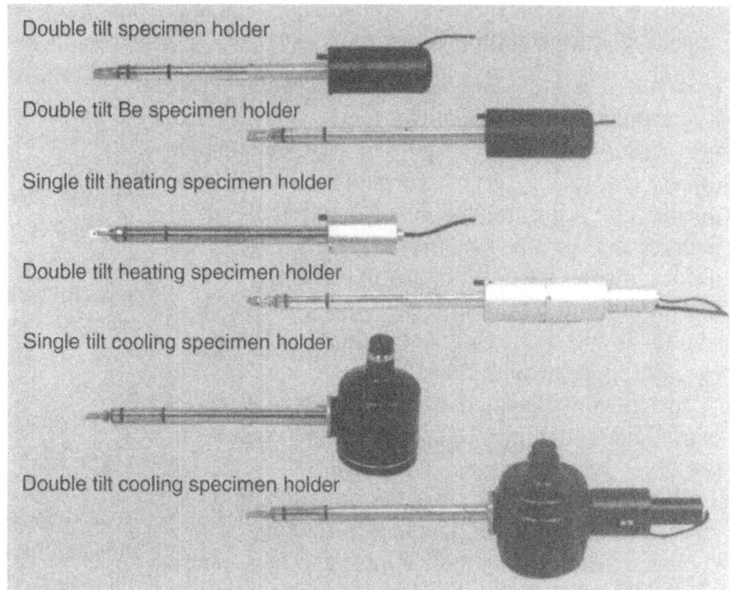

Fig. 2.14. Appearance of various side-entry type specimen holders

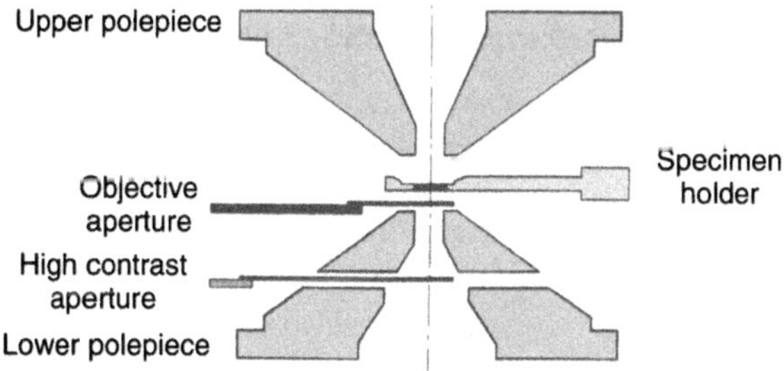

Fig. 2.15. Cross section of a polepiece of the objective lens

properties of the objective lens. Figure 2.15 shows a cross section of the polepiece of the objective lens in a conventional transmission electron microscope (JEM-2010F). A strong magnetic flux is generated in a space between the upper and lower polepieces. The specimen is set at the center of the polepieces, with an objective aperture located in a position below the specimen. The objective *stigmator* (not shown in Fig. 2.15) is installed below the lower polepiece. The four parameters discussed below are typical, indicating the main optical properties of the objective lens.

Focal Length. In a magnetic lens, stronger lens excitation generates a shorter focal length. The focal length (f_0) of the objective lens is almost constant because the objective lens is used under conditions of constant lens excitation. In general, a short focal length provides small spherical aberration and high resolution, whereas a longer focal length provides higher image contrast.

Spherical Aberration Coefficient. The point resolution (d) of a transmission electron microscope is determined by the spherical aberration coefficient (C_s) of the objective lens and the wavelength (λ) of the incident electrons.

$$d = 0.65(C_s\lambda^3)^{1/4} \tag{2.3}$$

In general, stronger lens excitation of the objective lens provides a smaller C_s, so it provides a higher point resolution.

Chromatic Aberration Coefficient. A variation of the current (ΔI) for objective lens excitation generates variation of the focal length and produces chromatic aberration. A variation of the accelerating voltage (ΔV_v), the energy spread of the electrons (ΔV_e) emitted from the filament, and the energy loss of the electrons (ΔV_s) due to inelastic scattering in a specimen generate variations of wavelength, resulting in chromatic aberration. Stronger lens excitation for the objective lens provides a smaller chromatic aberration coefficient (C_c). Variation of the focal length (Δ) due to chromatic aberration caused by these factors is given as:

$$\Delta = C_c\sqrt{\left(2\frac{\Delta I}{I}\right)^2+\left(\frac{\Delta V_v}{V}\right)^2+\left(\frac{\Delta V_e}{V}\right)^2+\left(\frac{\Delta V_s}{V}\right)^2} \tag{2.4}$$

where V is the accelerating voltage; and I is the objective lens excitation current.

Minimum Step of Defocus. A defocusing technique making the objective focus slightly underdefocused is widely used to obtain higher image contrast during transmission electron microscopy. Especially in high-resolution electron microscopy, a *through-focus method* (serially defocusing) is used. With this method the lower the minimum defocusing step (Δf) is set, the smaller is the defocusing series that can be used for obtaining through-focus images.

The objective lens generally produces astigmatism. Such astigmatism should be corrected with the objective stigmator.

The function of the objective lens forming an image results from the postfield of the lens at the backside of the specimen. On the other hand, a prefield works as a condenser lens in current analytical electron microscopes. Because of this strong prefield, a small electron beam probe can be used. The objective lens of this type, with simultaneous functions of the condenser lens and forming an image, is called the *condenser-objective* (C-O) lens.

An objective aperture is generally located at the back focal plane. The image contrast appears by using an objective aperture. Figure 2.16a shows the image-forming mechanism with an objective lens and an objective aperture. Some of the scattered electrons (diffracted waves) can be removed by an objective aperture, and the image contrast appears accordingly. It looks like absorption of the incident electrons inside the specimen. The image contrast formed by this mechanism is called *absorption-diffraction contrast* or *amplitude contrast*. Under tilted illumination conditions using deflectors, dark-field imaging can be carried out by selecting a specific diffracted wave by an objective aperture. The back focal plane of the objective lens corresponds to the reciprocal space. The objective aperture can limit some higher spatial frequencies, and therefore it should be carefully utilized for high-resolution electron microscopy. For high-resolution electron microscopy, the image contrast is created by interference of the transmitted and diffracted waves, as shown in Fig. 2.16b; it is called *phase contrast* [1].

2.1.5.2 Magnifying Lens System (Intermediate and Projector Lenses)

It is possible to change magnifications in a wide range (e.g., low magnification of about ×50 to magnifications of more than ×1 500 000) with the transmission electron microscope. It can be done by changing the focal length with changing the

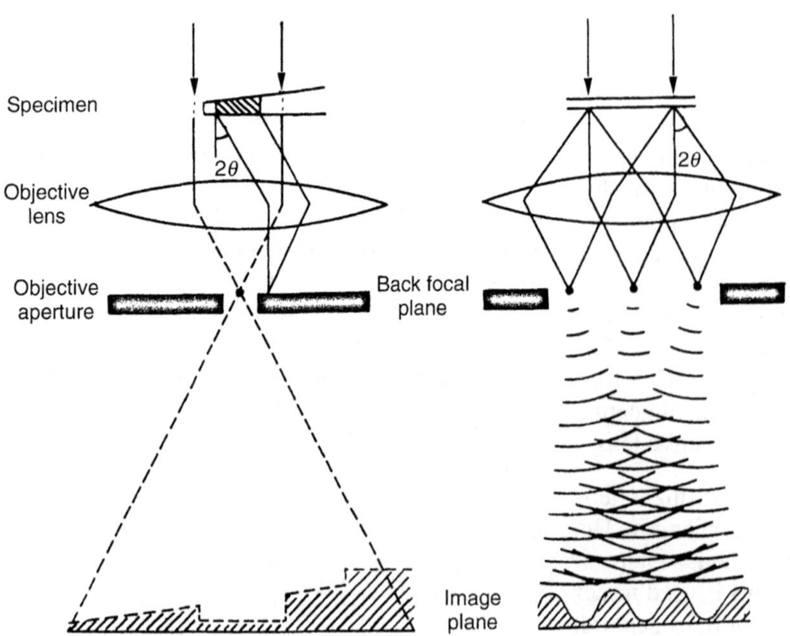

Fig. 2.16. Imaging process with an objective lens and the effect of the objective aperture. **a** Absorption-diffraction contrast (amplitude contrast). **b** Phase contrast. *Hatched areas* indicate the image intensity

Principle of Electron Lens Action

Magnetic flux generated by the lens coil is condensed at the tip of a polepiece through the yoke. The polepiece has the shape of rotational symmetry around the optical axis, with a bore diameter b and a gap distance S between the poles, as shown in Fig. 2.17a. It is designed such that the flux condenses at the gap.

The electrons, which pass along the optical axis exactly, do not suffer the Lorentz force due to the magnetic field. An incident electron at a distance r from the axis suffers the Lorentz force in a direction from back to front of the paper by an r-component B_{1r} of the magnetic field B_1. The Lorentz force is given by Eq. 5.13. The direction of the force is given by Fleming's left-hand rule. The electron starts to make a rotational motion clockwise but then experiences the r-direction force by z-component B_{2z} of the magnetic field B_2, which converges the electrons to the axis. Thus the electron is focused at a point on the axis (focal point).

Figure 2.17b shows an electron-focusing trajectory (the distance from the optical axis) neglecting the rotational motion. The focusing action of the magnetic lens can be considered to be a convex lens of light. Figure 2.17c shows the rotational trajectory of an electron. A rotation angle is almost proportional to the total magnetic flux of the lens. Thus the image formed by the magnetic lens is, in general, a rotated image, not the inverted image obtained with an optical lens.

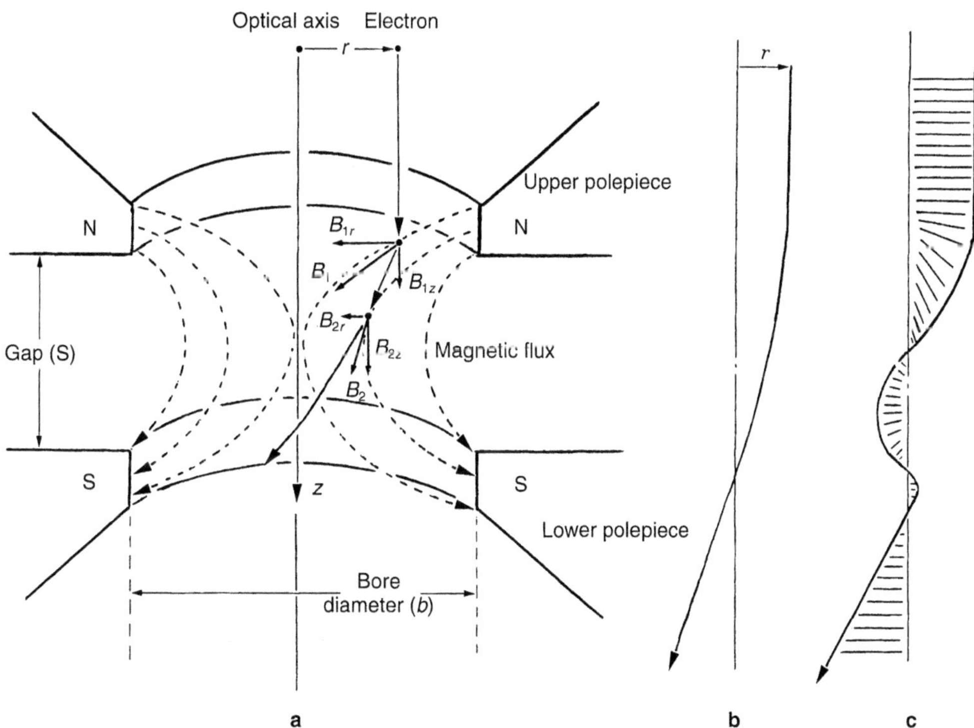

Fig. 2.17. Principle of the action in the electron lens. **a** Electron pass in the polepiece. **b** Electron focusing trajectory indicating the distance of the electron from the optical axis. **c** Rotational trajectory of an electron

strength of the magnetic field in the magnetic lens. The image-forming mechanism of the electron lens can be understood in the same way as geometrical optics with the optical lens. Figure 2.18b shows the mechanism for magnifying an image. A transmitted image of the specimen illuminated with electrons is formed and magnified by the objective lens at first. Then the image is magnified further through two to four stages with the magnifying lens system, which consists of an objective minilens (OM lens), an intermediate lens, and a projector lens; the image is projected on a phosphor screen or a photo-film. At extremely low magnification (so-called Low-Mag mode), the

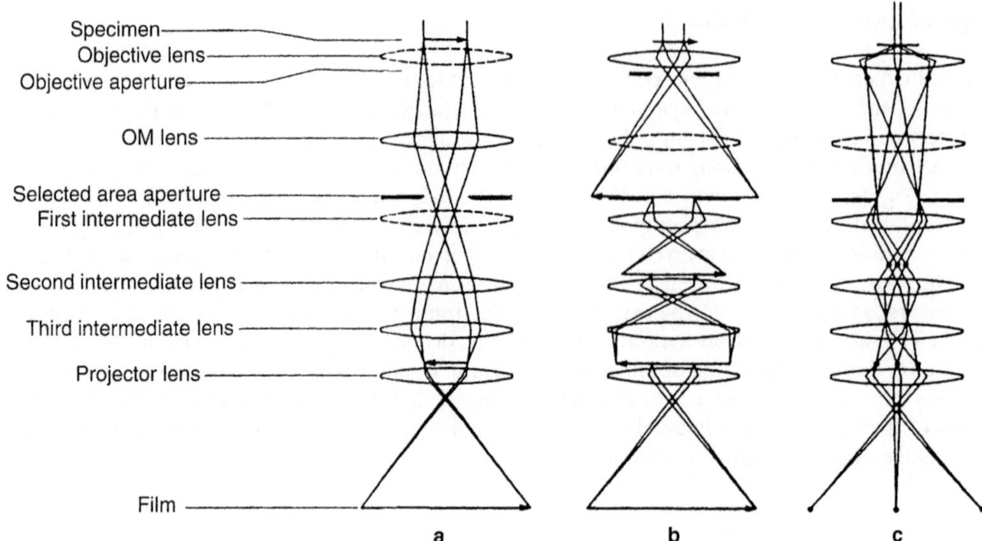

Fig. 2.18. Image-forming lens system. **a** Low-Mag mode. **b** Mag mode. **c** Diff mode

image is formed by the OM lens, the magnified intermediate lenses, and the projector lens without the objective lens, as shown in Fig. 2.18a. Also, the first intermediate lens is turned off in this case. The image rotates with the change of magnification, as the total magnetic flux strength of the lenses changes. Recently, an image rotation-free system, where the total flux strength is kept constant even if the magnification is changed, and also the image-orientation system, where total flux strength can be changed to keep the magnification constant, have been developed. The magnification is defined as the ratio of a specimen size and an image size recorded on a photo-film. It is calibrated at the film plane. To attain a magnification exactly as indicated in the instrument, the following points should be noted in the operation: (1) The specimen should be set accurately at the so-called Z-position in the objective lens. (When the objective lens is excited by the specific lens current condition, defined by its optical performance, the Z-position is defined by the just-focused position.) (2) The focus of the objective lens should be adjusted to be the just-focused position.

In general, the magnification of the image-forming lens system has an error of 5%–10%, so the final magnification in the instrument has an error of 5%–10%. To obtain higher accuracy during magnification, the magnification should be calibrated with a lattice image of a standard specimen (whose lattice spacing is known) by recording the specimen in the same field of view as the standard. In addition, the instruments may give a few percent image distortion, so the distortion should be calibrated by keeping data on the performance of the instrument.

Figure 2.18c shows the Diff mode, which provides an electron diffraction pattern. Like X-ray diffraction, electron diffraction is useful for determining the crystal orientation, lattice spacing, and so on. In both the Mag mode shown in Fig. 2.18b and the Diff mode shown in Fig. 2.18c, the OM lens is turned off. In the Mag mode, the focus of the first intermediate lens is adjusted to the image plane of the objective lens where a selected area aperture is inserted. In contrast, in the Diff mode the focus of the first intermediate lens is adjusted at the back focal plane of the objective lens. Therefore, the magnifying lens system magnifies an electron diffraction pattern formed on the back focal plane. Figure 2.19 shows the principle of electron diffraction and the mechanism magnifying the electron diffraction pattern in detail. The incident electrons are diffracted by the specimen, depending on its crystal structure. The diffraction angle 2θ is given by the Bragg condition as

$$2d\sin\theta = \lambda \tag{2.5}$$

where d is the lattice spacing of the specimen, and λ is the electron wavelength. The diffracted electrons form a diffraction pattern on the back focal plane via the objective lens. A camera length L_0 corresponds to the focal length f_0 of the objective lens. (The camera length is defined as a distance

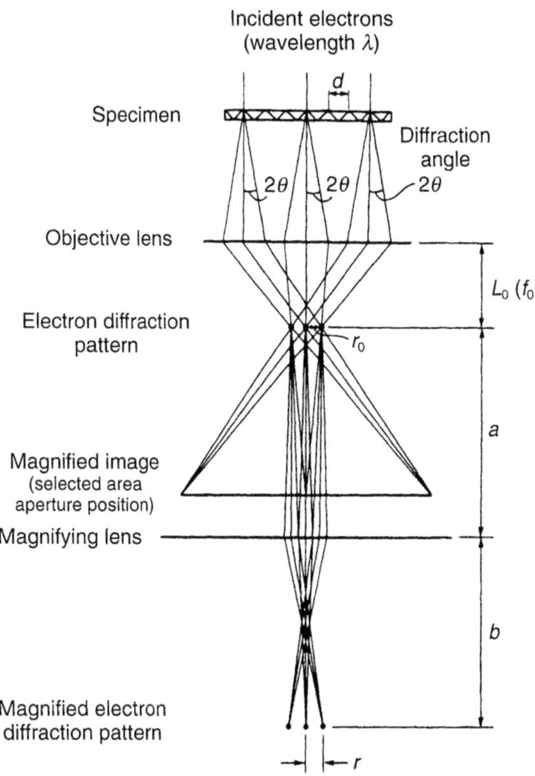

Fig. 2.19. Magnifying an electron diffraction pattern

from the specimen to the film plane when the lens is not used. Because the lens is used in a transmission electron microscope, the length may be called an effective camera length. However, it is commonly called simply a camera length.) The magnifying lens system magnifies the pattern and projects it onto the fluorescent screen or the photo-film at magnification M ($M = b/a$). Therefore, the final camera length L becomes

$$L = L_0 \cdot M \tag{2.6}$$

Lattice spacing d corresponding to a distance r on the pattern is represented as follows.

$$d = L \cdot \lambda / r \tag{2.7}$$

The camera length is calibrated at the film plane. The magnification and camera length projected on the fluorescent screen are slightly smaller than the calibrated ones because the screen is located at a higher position than the film. To obtain a camera length exactly as indicated in the instrument, the following points should be noted during the operation.

1. Specimen should be set exactly at the Z-position in the objective lens.
2. Specimen should be illuminated by a parallel beam.
3. Focus of the objective lens should be adjusted to the just-focused position.
4. Focus of the first intermediate (Diff focus) should be adjusted at the back focal plane of the objective lens.

The magnification of the image-forming lens system generally has an error of about 5%–10%, so the final camera length indicated in the instrument includes an error of about 5% to 10% as well. To obtain more accuracy, the camera length should be calibrated with the diffraction spots of a standard specimen whose lattice spacing is known, by recording the diffraction pattern of a specimen in the same condition as the standard. It is possible to obtain the accuracy to three effective figures by simultaneously recording the patterns of a specimen in the same field of view as a standard crystal (e.g., evaporated gold particles). Note that an instrument with the capability of high-resolution diffraction without the lens is commercially available.

Figure 2.20 shows a relation between the spread of diffraction spots and the convergence angle of the incident electron beam. As shown in Fig. 2.20a, when the specimen is illuminated by a parallel beam the diffraction spots become infinitely small in principle. On the other hand, as shown in Fig. 2.20b, when the specimen is illuminated with the convergent beam, the spots change to disks. The convergence angle α is obtained by comparing the radius of the disk with the diffraction angle 2θ. The diffraction pattern formed by the convergent beam is called the *convergent beam electron diffraction* (CBED) pattern (see Section 5.1.2).

2.1.6 Viewing Chamber and Camera Chamber (Image Recording System)

It is possible to observe electron microscopic images and diffraction patterns projected on the fluorescent screen in a viewing chamber. Usually, a binocular is installed on the viewing chamber, as it is useful for focusing images. The fluorescent screen is an aluminum plate coated with phosphor powder. Lead glass is used for a viewing window to shield hard X-rays generated in the column of a transmission electron microscope. The thickness of the lead glass increases with an increase in

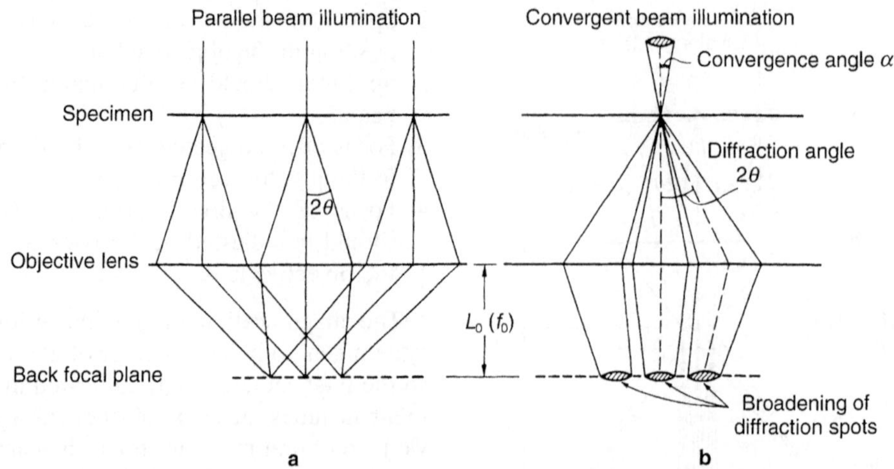

Fig. 2.20. Relation between illumination condition and broadening of diffraction spots. **a** Electron diffraction. **b** Convergent beam electron diffraction

accelerating voltage; therefore observing the fine image contrast on the screen generally becomes difficult in a high voltage electron microscope. In this case, a television (TV) camera installed below the camera chamber is useful.

The following recording systems for transmission electron microscopy are widely used these days.

2.1.6.1 Photo-film

Various sheet films are on the market, and such film is used widely as a conventional recording medium. Roll films of 70 and 35 mm are sometimes used from the viewpoint of economy. The film is exposed to electrons directly in the camera chamber of a transmission electron microscope. The sensitivity of the film for electrons is similar to that for light. The optical density D of film (negative) is obtained with the transmittance T of the film when it is illuminated with light as follows.

$$D = \log_{10}(1/T) \tag{2.8}$$

where

$$T = I/I_0 \tag{2.9}$$

where T is the transmittance of film; I_0 is the intensity of the illuminating light; and I is the intensity of transmitted light.

Higher D corresponds to higher density of film. Under the condition with $D > 1$, photo printing is difficult, so the film is usually exposed for satisfying D = 0.7–1.0. The relation between optical density D and exposure E is called the *characteristic curve*, or the *H-D curve* (after Hurter and Driffield, who first proposed it). The gamma (γ) showing the slope of the curve is called the contrast and is defined as:

$$\gamma = \tan\theta = \Delta D/\Delta \log_{10} E \tag{2.10}$$

A large γ value corresponds to high contrast, but it results in narrow latitude and thus a narrow dynamic range. Figure 2.21 shows characteristic curves for some films. It should be noted that the characteristic curve depends on the accelerating voltage and the conditions of development. Although resolution of the film depends on the type of film, it is around 10 μm in most cases.

Figure 2.22 shows a characteristic of the signal/noise ratio (S/N). This characteristic of the film is compared with that of the imaging plate (UR-III type; 50 μm/pixel), noted below. The characteristics of a high gain mode for the imaging plate and a low gain mode for the imaging plate reader are shown.

2.1.6.2 Television Camera

Observing and recording electron microscope images by a television (TV) system is convenient for in situ observation, multiuser observation, computer input, and instant printing. The TV observation of electron microsope images is carried out by a TV camera combined with a dedicated fluorescent screen. Figure 2.23 shows the constitution of detection units of TV cameras. The electron microscopic image is converted to a light image by the fluorescent screen of a transmission

Fig. 2.21. Characteristic curves of various kinds of photo-film (200 kV)

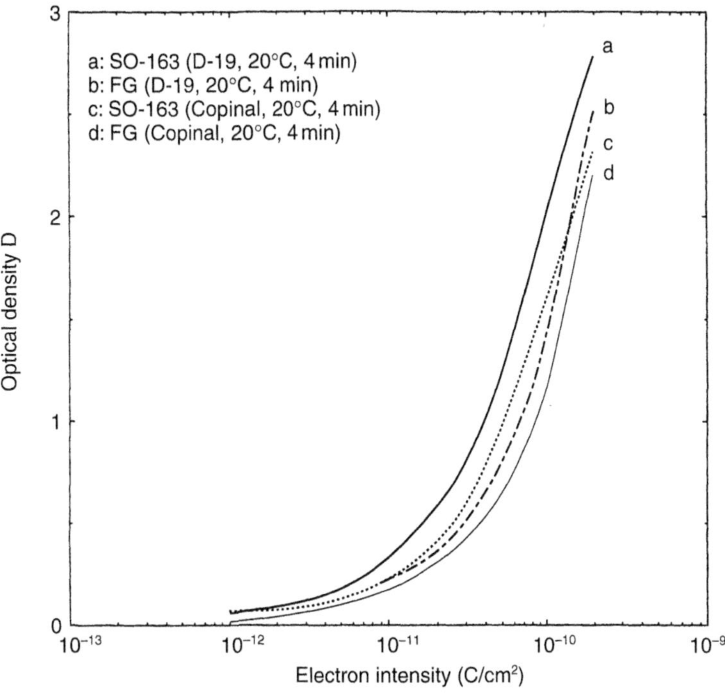

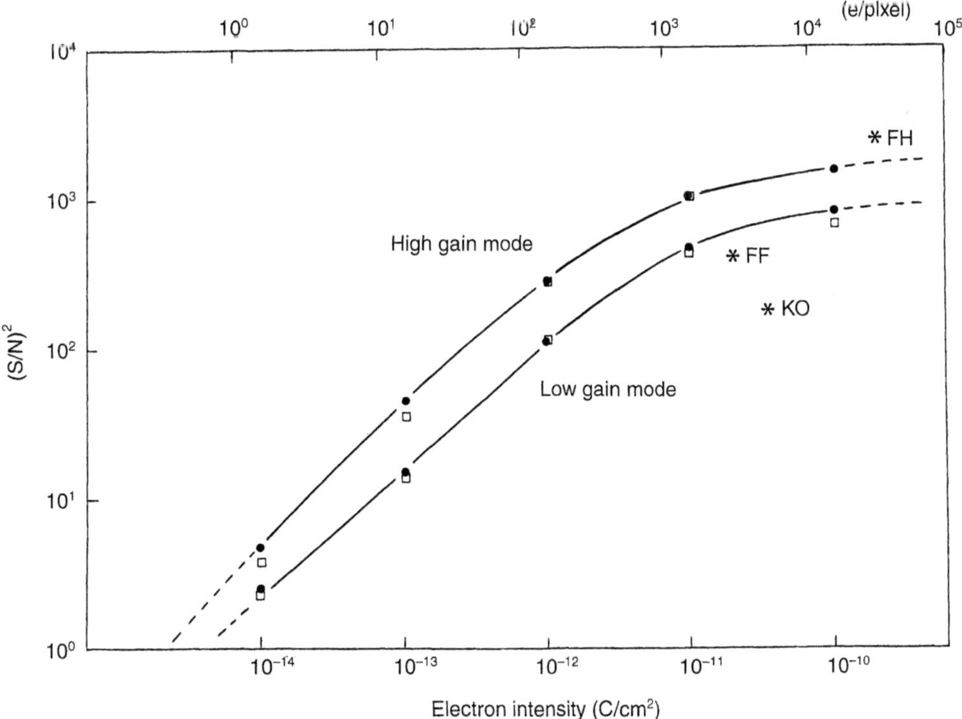

Fig. 2.22. The signal/moise ratio [$(S/N)^2$] of photo-film (*) and imaging plate (*squares, circles*). *, *Squares*, 100 kV; *circles*, 200 kV; *FH*, Fuji HR II; *FF*, Fuji, FG; *KO*, Kodak 4489

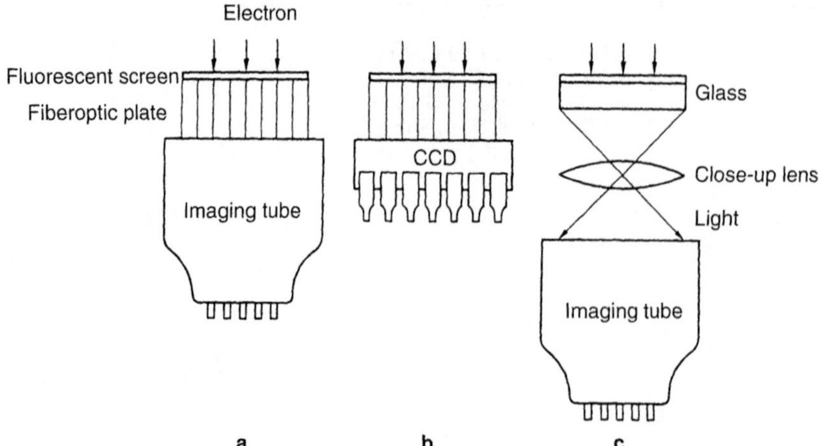

Fig. 2.23. Structure of detecting parts in TV cameras. **a** Fiber plate/imaging tube system. **b** Fiberoptic plate/Charge-coupled device (CCD) system. **c** Optical lens/imaging tube system

type; it is then transferred to a recording device through a fiberoptic plate or a close-up lens. The resolution of the fluorescent screen is about 100 μm, which depends on the kind of phosphor, its painting method, and the accelerating voltage. The fiber plate is composed of many 6 μm fibers. There is little loss of light intensity, and little image distortion is generated in the fiberoptic plate. On the other hand, blemishes of the fiber and contact patterns of the fibers sometimes appear in the image and are technical problems yet to be solved. The disadvantage is the small area of the detector, so the field of view in a TV camera is limited. The microlens can adjust the area of the view by changing the magnification, but then light absorption and image distortion appear. A silicon intensifier tube (SIT) was formerly used widely as a camera head. It accumulated and intensified images and therefore worked under low light level (e.g., a few millilux). It had the highest sensitivity in the imaging tubes. It was easy to operate because there was a wide dynamic range of the SIT for brightness. On the other hand, shading appeared (lack of image intensity at outer parts of the image) that was a typical property of the imaging tube. The SIT had been produced by handcrafted glass fabrication, but the market supply was recently stopped. An *image intensifier* can be used in a TV camera, with the imaging tube having lower sensitivity than the SIT, to increase the sensitivity. The TV camera with the intensifier has almost the same sensitivity as the SIT.

A charge-coupled device (CCD) has no shading, and it is compact and inexpensive. It has the disadvantages, however, of low sensitivity and a narrow dynamic range for electron intensity.

A video signal generated by a TV camera can be sent to a video signal processor, and then image processing (i.e., contrast enhancement, shading correction, image accumulation) is carried out. A process of image subtraction of [(image intensity)—(electron intensity without a specimen)] is especially useful for suppressing various blemishes produced in the system.

It is easy to transfer a TV image to a computer. Most personal computers can be installed in a video capture board with an input terminal of RS-170 or NTSC. The video cable should be connected to the video terminal of the board. It should be noted that some scanning microscopes have no standard video signal.

2.1.6.3 Imaging Plate

The imaging plate (IP) was originally developed as a high-sensitivity recording device for X-ray image [2]. The imaging plate for transmission electron microscopy was developed later [3,4]. The principle of the imaging plate for transmission electron microscopy is similar to that for X-ray imaging. It is a two-dimensional image recording device containing the photo-stimulatable phosphor—Eu^{2+} is doped in the phosphor halide [$BaF(Br,I){:}\,Eu^{2+}$] powder coated on a plastic base. The imaging plate (99.6 × 80.9 mm) for transmission electron microscopes is smaller than that for X-rays. Blue pigment is doped in the phosphor of the imaging plate for transmission electron microscopes, and so the imaging plate looks blue. The structure of such an imaging plate is shown in Fig. 2.24. The incident electrons generate hole-electron pairs in the phosphor. The generated electrons are then trapped at the defect position (lattice vacancies of negative ions) of the phosphor, and the holes are trapped at Eu^{2+}. When the laser beam is illuminated on the imaging plate, trapped electrons are released to the conduction band and are combined with holes; eventually they generate light. The light is amplified with a photomultiplier tube (PMT), converted to an electrical signal, and then reconstructed into images. After reading out the image data in the imaging plate reader, electron-hole pairs left in the plate are erased by exposing the plate to high-intensity light; an imaging plate can therefore be reusable (Fig. 2.25).

Figure 2.26 shows the relation between the output signal intensity and the number of incident electrons; and for comparison, the optical density of conventional photo-film is shown. Figure 2.26 shows that an imaging plate has higher sensitivity and a wider dynamic range than the film, as well as good linearity between the input and output signals. The data suggest the possibility of image recording under the condition of low electron intensity. It must be noted that the S/N ratio should be carefully taken into account to carry out image analysis accurately. Figure 2.27 shows the $(S/N)^2$ of an imaging plate as a function of the number of the incident electrons for 100 and 200 kV [5]. The $(S/N)^2$ increases with an increase

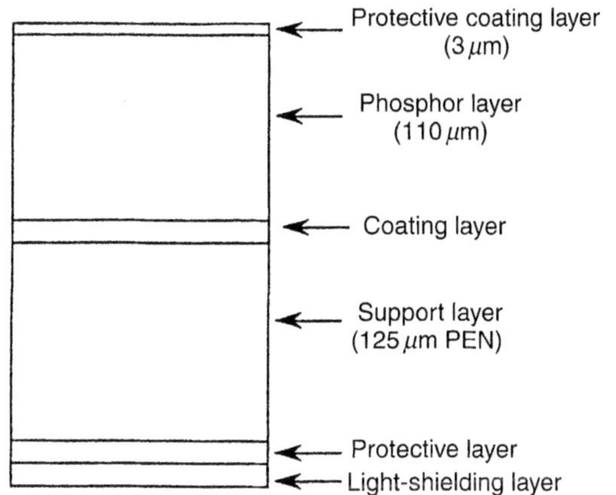

Fig. 2.24. Cross section of the imaging plate (FDL-UR-V) showing its constituent parts. *PEN*, polyethylene naphthalate

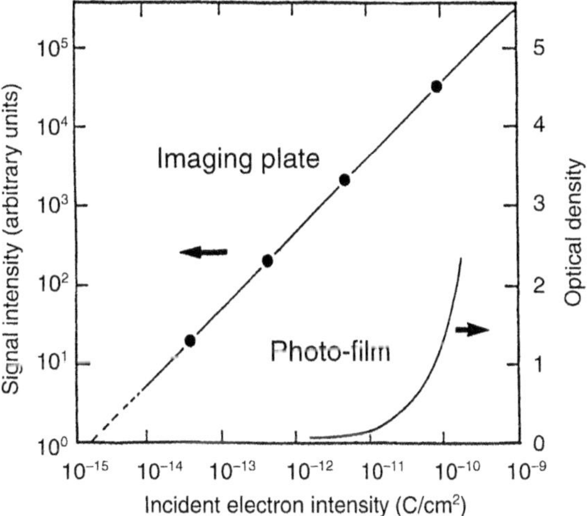

Fig. 2.26. Sensitivities of the imaging plate and photo-film for electron irradiation

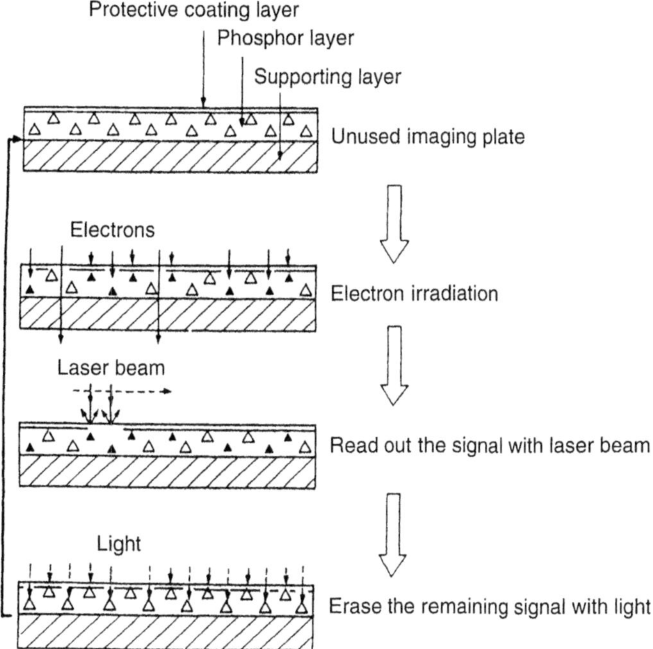

Fig. 2.25. Principles of recording, reading, and erasing processes on the imaging plate

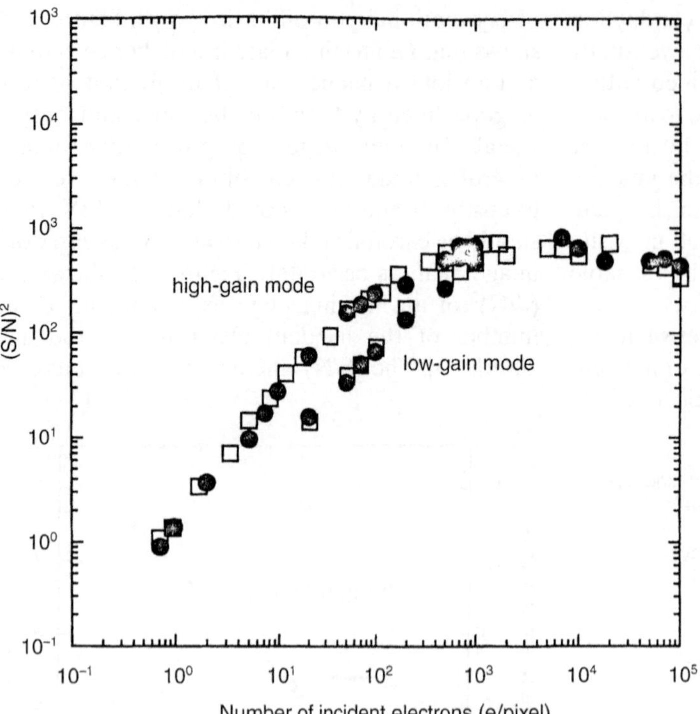

Fig. 2.27. The $(S/N)^2$ characteristics of the imaging plate (FDL-UR-V) for high-gain mode and low-gain mode (*squares*, 100 kV; *filled circles*, 200 kV)

in electron intensity. In a low electron intensity region the noise is relatively high, so it is difficult to carry out the image analysis with high precision. In regions of electron intensity higher than 10^2 electrons/pixel, the $(S/N)^2$ tends to be saturated. Thus, it is impossible to increase the $(S/N)^2$ even if the intensity of incident electrons is increased.

Detective quantum efficiency (*DQE*) is sometimes used as a parameter to indicate the performance of detectors. The DQE is defined using the S/N of the input signal $[(S/N)_{in}]$ and S/N of the output signal $[(S/N)_{out}]$ as follows.

$$\mathrm{DQE} = (S/N)_{out}^2 / (S/N)_{in}^2 \tag{2.11}$$

For the ideal recording system, the S/N of the output signal is kept the same as that of the input signal so the DQE is equal to 1. For evaluating the DQE, an effective pixel size of an imaging plate and thus a point spread function showing the spread of the signal in the detector should be evaluated. Figure 2.28 shows a point spread function measured for an imaging plate [5]. Figure 2.29 shows the DQE of an imaging plate for a high gain mode of the imaging plate reader, taking into account the point spread function. In the regions of low and high electron intensity, DQE has low values [5,6]. With low DQE, the S/N of the output

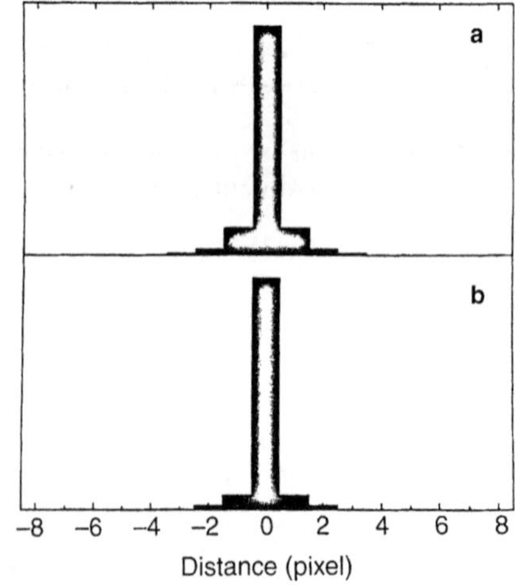

Fig. 2.28. Point spread functions of the imaging plate. **a** High-gain mode. **b** Low-gain mode

signal is low compared with the S/N of the input signal. Therefore, it should be noted that the high image quality with increased electron intensity is not expected under this condition. Thus, to observe beam-sensitive specimens, the electron intensity corresponding to the high DQE region is effective. In contrast, to observe beam-insensitive specimens, the electron intensity corresponding to

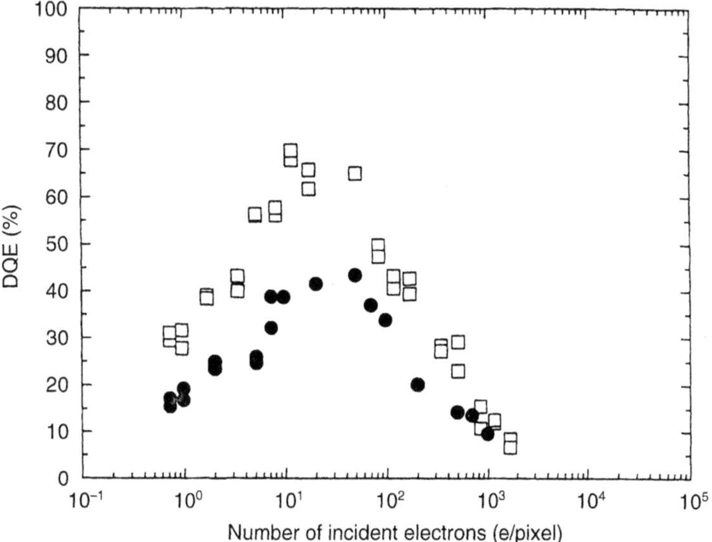

Fig. 2.29. Detective quantum efficiency (*DQE*) characteristic of the imaging plate for electron irradiation. *Squares*, 100 kV; *filled circles*, 200 kV

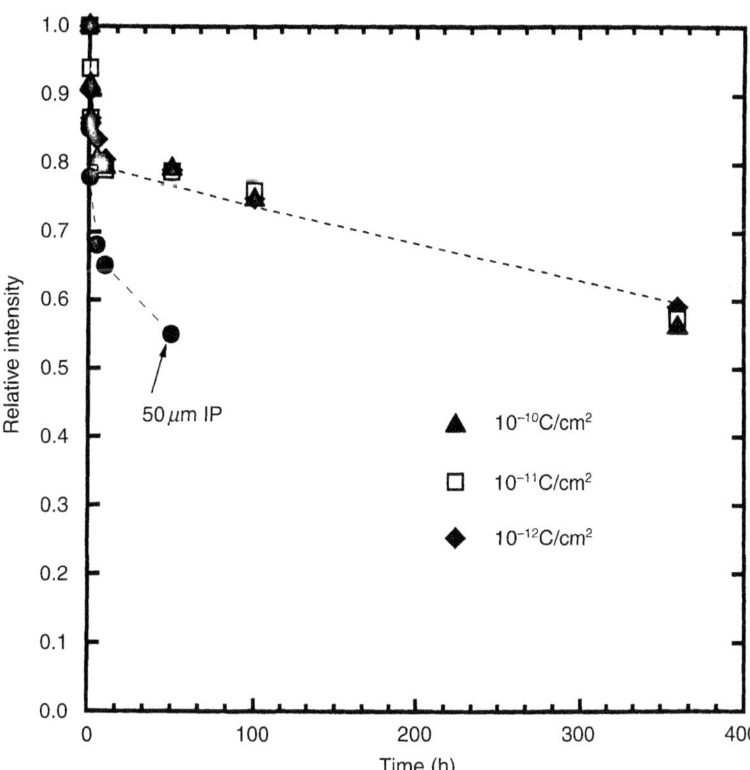

Fig. 2.30. Fading characteristic of the imaging plate (FDL-UR-V). 50 μm IP indicates UR-III type

a high S/N is appropriate. It is known that the DQE of an imaging plate generally decreases with an increase in accelerating voltage. This is because the penetration power of the incident electrons in the phosphor layer increases with the increase in accelerating voltage, and so some of the electrons pass through the phosphor layer. The tendency for DQE to decrease on an imaging plate in regions of low and high electron intensity is basically the same as for a slow-scan CCD camera [7–9].

Imaging plates show the fading phenomenon, which is the fading of recorded signals with time after exposure. Figure 2.30 shows variations of output signal intensity as a function of the passage of time after the exposure [10]. The curves show that fading is especially pronounced for the first 5 h after electron exposure. In general, the signal intensity $I(t)$ of the imaging plate at $(t+0.1)$ h after electron exposure is given by:

$$I(t) = I_0[c_1 \cdot \exp(-0.638 \cdot t/T_1) + c_2 \cdot \exp(-0.638 \cdot t/T_2)] \quad (2.12)$$

where I_0 is the signal intensity at 0.1 h (6 min) after electron exposure, which may be the shortest time it takes to read the imaging plate data; c_1 and c_2 are constants with the relation $c_1 + c_2 = 1$; and T_1 and T_2 may correspond to half-value periods of a well-known formula for the disintegration of radioisotope elements. These constants for the imaging plate of 25 μm/pixel (FDL-UR-V type) at 25°C are as follows [10]:

$$c_1 = 0.2, \quad c_2 = 0.8$$
$$T_1 = 1.2\,h, \quad T_2 = 850\,h \quad (2.13)$$

and those for the imaging plate of 50 μm/pixel (UR-III type) are as follows [11]:

$$c_1 = 0.3, \quad c_2 = 0.7$$
$$T_1 = 2\,h, \quad T_2 = 170\,h \quad (2.14)$$

It has also been reported that the fading phenomenon is less pronounced at lower temperatures (e.g., 0°C) [11]. As the fading phenomenon is related to the passage of time t (Eq. 2.12), the intensity of the output signal is proportional to the incident electron intensity; thus quantitative analysis can be carried out using the intensity of the output signal. Furthermore, by using a sensitivity curve such as that shown in Fig. 2.26, which gives the relation of the incident electron intensity and the output signal intensity over time, the absolute value of the incident electron intensity can also be obtained from the output signal intensity.

2.1.6.4 Slow-Scan Charge-Coupled Device Camera

Figure 2.31 shows the constitution of a slow-scan charge-coupled device (CCD) camera [12] used in transmission electron microscopes. Incident electrons are converted to light by the YAG (yttrium-aluminum-garnet: $3Y_2O_3 \cdot 5Al_2O_3$) scintillator and transferred to the CCD through the fiberoptic plate. The light detected is converted to an electron charge that is temporarily stored in each channel at the semiconductor electrode on the surface of the CCD. The accumulated electrical charge is sequentially transferred to the neighboring pixel while sweeping and then sent out from a terminal as the electrical signal. Thus, by storing the electrical charge for some time and reading it out with scanning, a slow-scan CCD camera has a higher sensitivity and a wider dynamic range than a real-time CCD camera. The dark current producing the noise and the background on the image can be reduced by cooling the CCD. Figures 2.32 and 2.33 show the S/N and DQE characteristics of a slow-scan CCD camera (Gatan model 792 BioScan) at 100 kV [9]. It takes

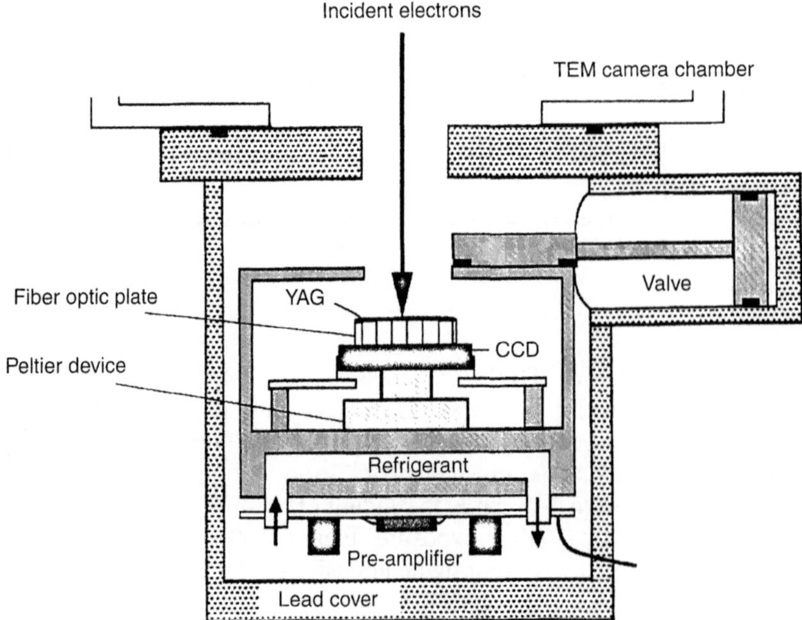

Fig. 2.31. Constitution of a slow-scan CCD camera

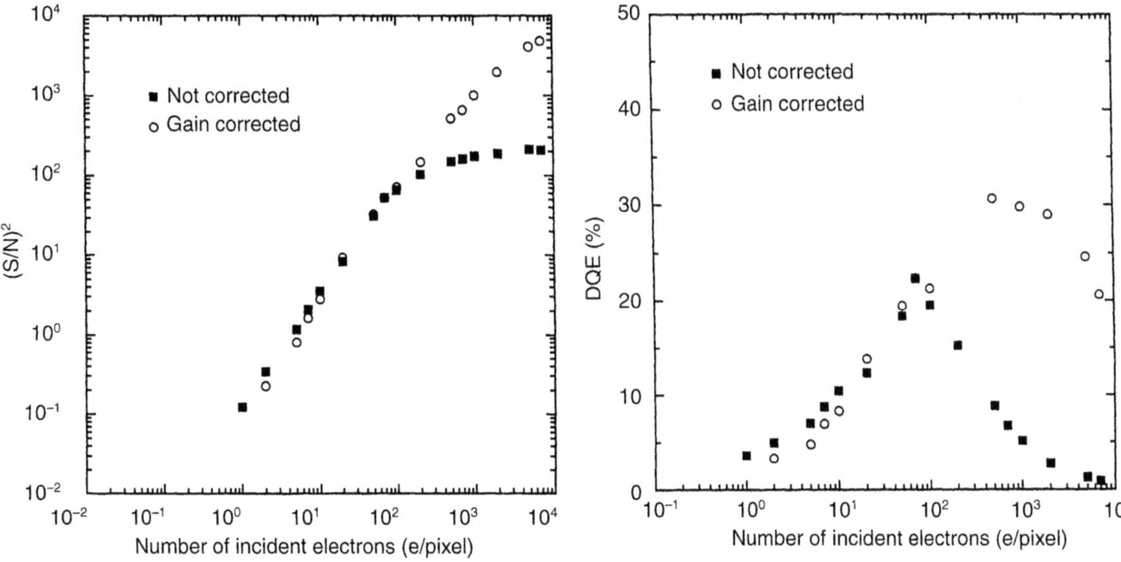

Fig. 2.32. $(S/N)^2$ of a slow-scan CCD camera

Fig. 2.33. DQE characteristic of a slow-scan CCD camera

Table 2.2. Comparison of performances of imaging plates and slow-scan CCD cameras.

Pixel size	Number of pixels	Dynamic range	Advantage	Notes
Imaging plates				
25 × 25 μm (50 × 50 μm)	3000 × 3760 (2048 × 1536)	4–5 figures	Possible for various kinds of electron microscope	Fading characteristics
Slow-scan CCD				
24 × 24 μm	1024 × 1024 (2048 × 2048)	4 figures	Digitized image can be seen in a few seconds	Artifacts with too strong electron intensity

Pixel sizes and pixel numbers in the typical systems are presented. The sizes and numbers in other systems are indicated in parentheses

only a few seconds to display the digital image data on a monitor after electron exposure in a slow-scan CCD camera. Table 2.2 compares the main characteristics of imaging plates and slow-scan CCD cameras.

2.2 Operation of Transmission Electron Microscopes

2.2.1 Alignment of Lenses and Astigmatism Correction

Microscope adjustments in daily operation are mentioned below. The order of the adjustments listed below follows the order of the procedures. It is a basic rule of microscope operation that the adjustments be carried out from upper parts to lower ones. During the adjustment procedure the objective excitation current or voltage (objective focus) should be set at the original value ("focus current" or "focus voltage"[1]), and the specimen position should be set as the original Z-position. Before starting the adjustments, two points mentioned above should be definitely confirmed. When some of the parts deviate largely from the optimum condition, it is difficult to adjust one part accurately. In this case, roughly adjust all parts and then repeat the adjustment for all parts, finally making the adjustments accurate, step by step.

[1] In electron microscopes the original value is sometimes indicated by the lens excitation voltage instead of the lens excitation current when they are related by the resistance in the circuit.

1. Adjust the electron gun.
2. Align the condenser lenses.
3. Adjust the astigmatism in the condenser lens.
4. Align the voltage center of the objective lens.
5. Adjust the astigmatism in the objective lens.
6. Adjust the astigmatism in the intermediate lens.
7. Align the projector lens.

2.2.1.1 Adjustment of an Electron Gun

The method for adjusting a thermionic emission gun is described.

Adjusting Bias Voltage and Filament Temperature. In general, an emission current increases with the increase in bias voltage, but the heating temperature necessary for emission saturation is higher. The emission current increases with the increase in heating temperature, and the energy spread of electrons emitted widens. The lifetime of the filament shortens with higher heating temperature. With the usual conditions for an LaB_6 filament, an emission current of about $15\mu A$ should be used. The practical adjustment procedures are as follows.

1. Observe the filament pattern projected on the screen at the standard heating temperature.
2. Set the emission current for about $15\mu A$ at the saturated condition by adjusting both heating temperature and bias voltage. The bias voltage depends on the mechanical position of the filament.
3. Set a stopper on the heating knob at the final position after determining the condition.

Alignment of an Electron Gun. Make a focused beam in the image mode without a specimen. Next, create an unsaturated condition of the filament by slightly lowering the heating temperature. A detail of a filament pattern can now be observed on the screen. The pattern should be adjusted so it is symmetrical (fourfold symmetry for an LaB_6 filament) by adjusting the tilt knobs (X and Y) of the gun alignment. For a normal filament the beam intensity becomes maximum for a symmetrical pattern. Figure 2.34 shows a typical emission pattern of an LaB_6 filament in the unsaturated condition. When the shape of the filament changes, maximum intensity is attained for an asymmetrical pattern. In this case, make a compromise using moderate conditions (fairly high intensity) and a pattern that does not cause a drastic loss of symmetry. When the shape of the filament is severely deformed, replace it with a new one. After finishing the adjustment, the heating condition should be changed to that for the saturated condition.

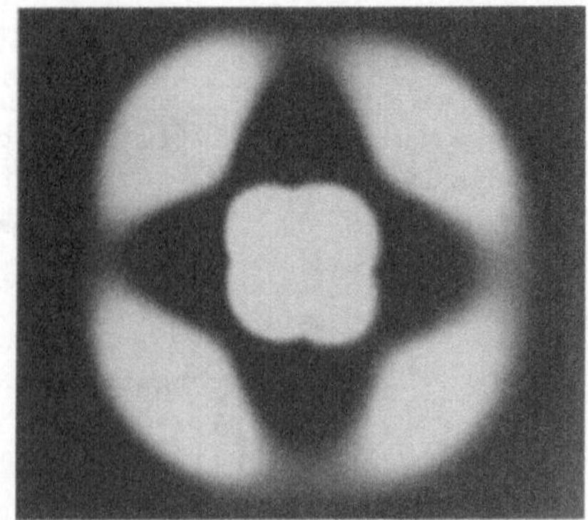

Fig. 2.34. Filament pattern of LaB_6 under unsaturated conditions

In the case of a thermal FEG, turn on an anode wobbler and adjust the alignment of the axis using the tilt knobs (X and Y) for gun alignment to stop the movement of the beam in the center.

2.2.1.2 Alignment of the Condenser Lenses

Select a large spot size (first condenser lens weakly excited) with the spot size knob and adjust the beam to the center of the screen using the gun alignment shift knobs (X and Y). Next, select a small spot size (first condenser lens strongly excited) with the spot size knob and adjust the beam to the center of the screen using condenser alignment shift knobs (X and Y). These procedures should be repeated a few times to find the beam on the center on the screen without a shift, irrespective of the spot size.

2.2.1.3 Adjustment of Astigmatism in the Condenser Lens

Focus the beam on the screen using the brightness knob (focusing in the second or third condenser lens). Make the shape of the focused beam circular by adjusting the condenser stigmator (X and Y) of the second or third condenser lens. With this procedure it is easy to observe astigmatism by changing the focus of the condenser lens from over to under (or under to over) by the brightness knob. With this operation, if the beam is defocused

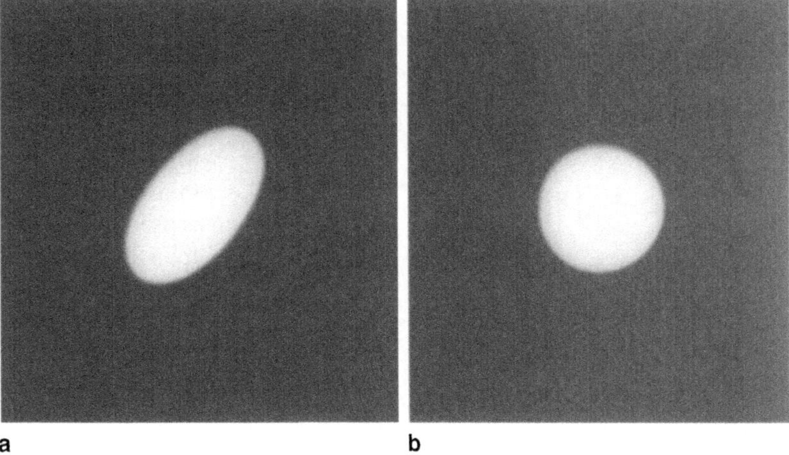

Fig. 2.35. Adjustment of astigmatism of a condenser lens. **a** Astigmatism exists. **b** No astigmatism

with a circular shape, there is no astigmatism. If there is astigmatism, the beam forms an elliptical shape owing to the change of the beam focus, as shown in Fig. 2.35. The astigmatism of the condenser lens depends on the condenser aperture size and spot size, so it should be adjusted when the aperture size or the spot size (or both) are changed.

2.2.1.4 Alignment of Voltage Center of the Objective Lens

A high voltage wobbler is used to align the voltage center of the objective lens. In the image observation mode, turn on the high voltage wobbler. The images, which are then magnified and demagnified continuously, can be seen because of the change in the high voltage. The center of the movement from expansion to contraction (or contraction to expansion) is the high voltage center. The center should be adjusted to the screen center by operating the deflectors (tilt X and Y of condenser alignment) (Fig. 2.36). The objective aperture should be retracted from the optical axis during this operation, and it should be inserted after the adjustment. Use of an image with a triple point in the carbon microgrid is convenient for this adjustment, as shown in Fig. 2.37. Adjust the wobbler center to the center of the screen with the tilt knobs of the double deflector system, as the image has only the distortion without the shift, as shown by the arrows in Fig. 2.37. The high voltage center depends on the spot size and magnification, so it should be adjusted when these conditions change. The dependence is especially remarkable for electron microscopes with a C-O lens producing a strong magnetic prefield.

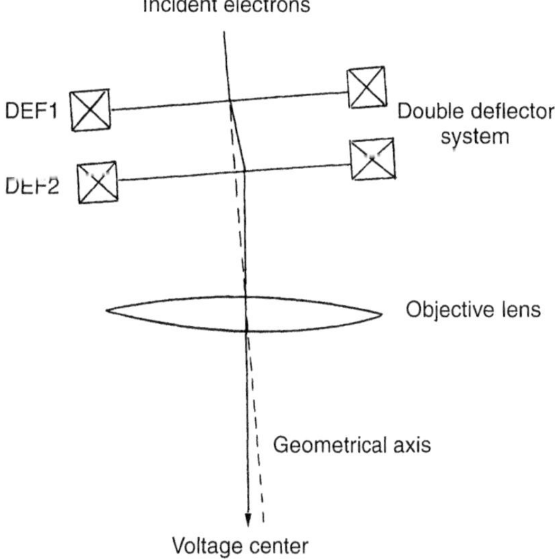

Fig. 2.36. Principles of adjusting the voltage center

Current center alignment is another way to align the objective lens. The objective wobbler is used to change the objective lens current. However, recent instruments have little instability of the objective lens current, so image broadening is mainly caused by wavelength change due to inelastic scattering of electrons in a specimen. Therefore, voltage center alignment is more widely used than objective lens alignment today.

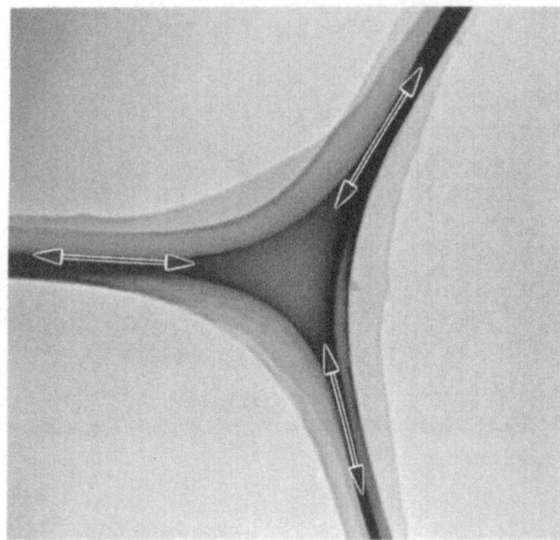

Fig. 2.37. Triple point in microgrid used to adjust the voltage center

2.2.1.5 Astigmatism Adjustment of the Objective Lens

Astigmatism of the objective lens is adjusted with the objective aperture inserted. Because astigmatism depends on the aperture size and its position, the astigmatism must be corrected each time conditions change.

Astigmatism Adjustment in the Low and Middle Magnification Range. The circular or square hole in a specimen (a carbon microgrid is convenient) is used for the adjustment at magnifications lower than ×100 000. In the just-focused condition, adjust the objective stigmator (X and Y) to obtain the same image contrast of the hole for two directions perpendicular to each other at an edge. Then make it underfocused and be sure the image contrast and the width of fringes at the hole edge in the two directions are the same. Again, make it just focused and adjust the stigmator as above. Next, go to the overfocused condition and make sure that the overfocused fringes are the same as for the underfocused condition. If necessary, repeat a few times.

Astigmatism Adjustment in High Magnification Range. For magnifications higher than ×200 000, an amorphous region in the specimen or a carbon-contaminated region at the edge of the specimen should be found. The thin amorphous carbon film is suitable for the astigmatism adjustment (as shown in Fig. 2.39).

High Voltage Wobbler and Image Wobbler

The high voltage wobbler is a function of adding about ±250 V to the accelerating voltage by changing the standard voltage periodically on the high voltage generator. The optical axis for the high voltage drift is called the high voltage center, and it does not coincide with the geometric axis. The deviation is thought to result from such factors as the magnetic property of the polepiece and the mechanical precision of the fabrication. Aligning the incident electron beam to the high voltage center is called alignment of the voltage center. The high voltage wobbler is used for this operation. In the transmission electron microscope, the objective lens focuses electrons with various energies, as if the accelerating voltage were changed. This is because the incident electrons suffer energy loss in the specimen. Hence, aligning the voltage center is important for obtaining high quality images.

Image wobbler is a function of tilting an electron beam on the just-focused position of the specimen by the double deflection system shown in Fig. 2.38. The image wobbler is used for setting the just-focused position of the specimen and for obtaining the just-focused condition of the objective lens.

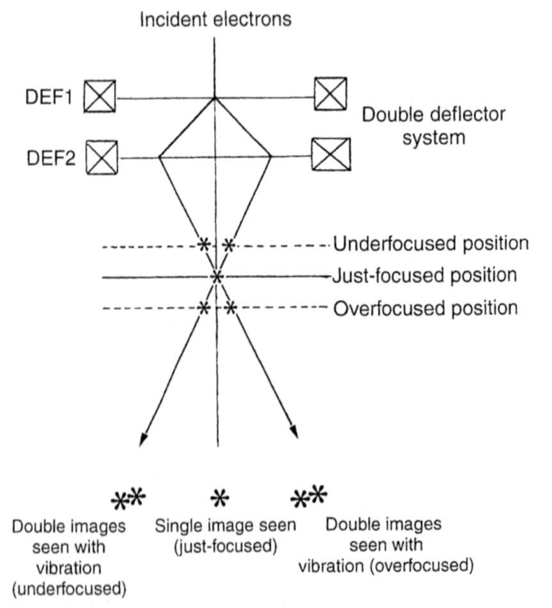

Fig. 2.38. Principles of image wobbler

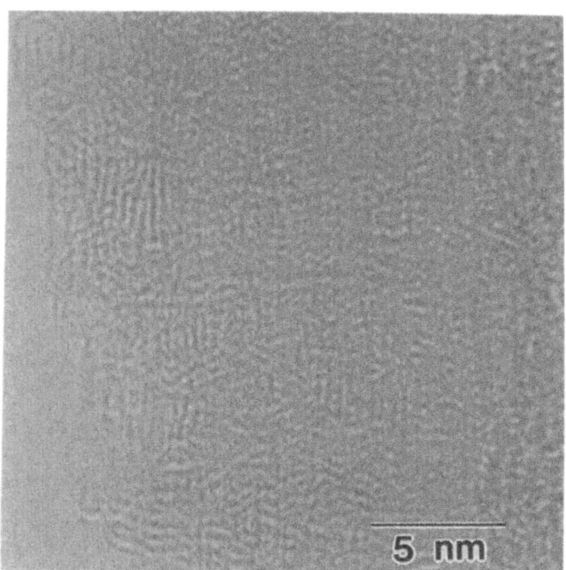

Fig. 2.39. Image contrast of amorphous film showing astigmatism of the objective lens

First, adjust the objective stigmator (X and Y) to remove a unidirectional pattern in the phase contrast image. Next, make it underfocused and be sure that the unidirectional pattern does not appear on a granular contrast image. Again, take it to the just-focused condition and adjust the stigmator as noted above. Next, go to the overfocused condition and be sure the granular image contrast without the unidirectional pattern is the same as for the underfocused condition. If necessary, repeat the procedures a few times. Finally, confirm the minimum image contrast without the unidirectional pattern at the just-focused condition.

2.2.1.6 Astigmatism Adjustment of the Intermediate Lens

Remove an objective aperture and insert a selected area aperture. Obtain an electron diffraction pattern of a specimen on the screen. Adjust the diffraction focus by controlling the first intermediate lens (use a "DIFF-focus" knob). Adjust the intermediate lens stigmator (X and Y) until the shape of the direct beam is circular. For this operation, it is easy to observe the magnitude of the astigmatism by changing the diffraction focus from over to under (or under to over).

2.2.1.7 Alignment of the Projector Lens

With a diffraction pattern on the screen, adjust the projector lens deflector (X and Y) to obtain a transmitted beam of a diffraction pattern at the center of the screen. Misalignment of the projection lens does not have a major influence on image quality, so the deflection can be used for the purpose of moving a specific part of the projected image to be investigated to the center of the screen. Adjusting the image center for an off-axis TV camera is one of its applications.

2.2.2 Focus Adjustment of the Objective Lens

2.2.2.1 Adjustment at Low and Middle Magnification

There are some basic points to be kept in mind for electron microscope observations; for example, the specimen position is set just at the Z-position, and the objective lens excitation is set at the just-focused condition. The focus adjustment using the image wobbler is also a basic operation.

First, adjust the objective lens excitation current to the just-focused condition (so-called focus current or focus voltage; information on the specific value for each instrument is available in the operating manual or from service engineers). Observe the image of a specimen, and turn on the image wobbler switch. The image vibration is sometimes observed at this point. Image vibration indicates that the specimen position deviates from the exact Z-position. Adjust the specimen position (height) to stop the vibration.

If vibration of the image is observed when the specimen is at the exact Z-position, the objective lens current may be deviating from the focus-current condition. Adjust the lens excitation with the focus knob to stop the vibration. It is difficult to distinguish overfocusing from underfocusing by image vibration alone. Change the focus or the Z-position and watch the change in image movement carefully to identify the just-focused condition.

2.2.2.2 Adjustment at Higher Magnification

At magnifications higher than ×200 000, the image contrast or Fresnel fringes at a specimen edge can be used for focus adjustment. The dark image contrast of Fresnel fringes appears in the overfocused condition, and a fringe with bright image contrast appears in the underfocused condition. In the just-focused condition, the fringes at the edge disappear and image contrast at a thin area be-

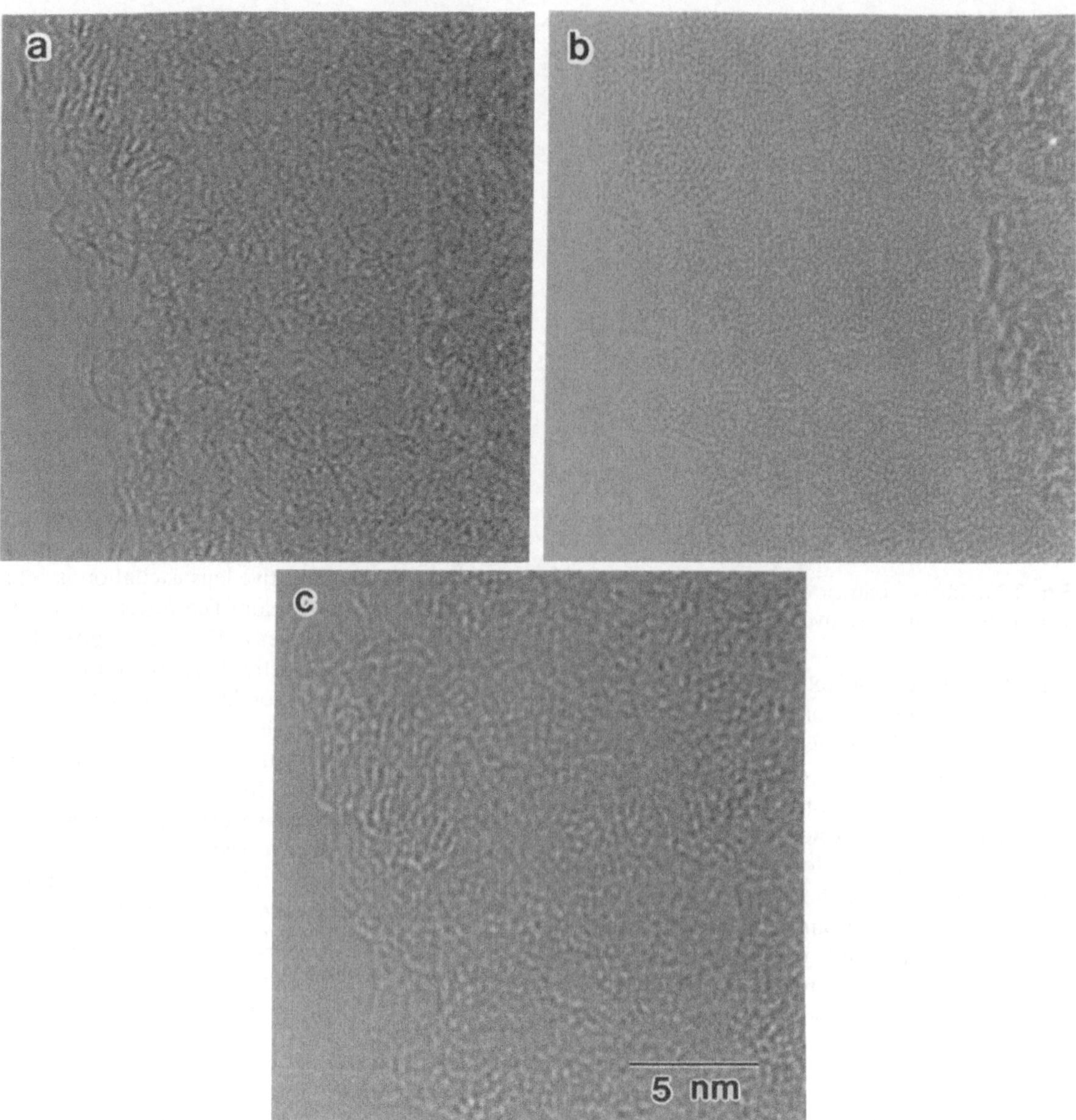

Fig. 2.40. Image contrast of amorphous film. **a** Underfocused. **b** Just-focused. **c** Overfocused

comes minimum, as shown in Fig. 2.40. For high-resolution electron microscope observations that are slightly underfocused, the so-called Scherzer focus is widely used as the optimum focus condition [1].

References

1. Shindo D, Hiraga K (1998) High-resolution electron microscopy for materials science. Springer, Tokyo
2. Sonoda M, Takano M, Miyahara J, Kato H (1983) Computed radiography utilizing scanning laser stimulated luminescence. Radiology 148:833
3. Mori N, Oikawa T, Katoh T, Miyahara J, Harada Y (1988) Application of the "imaging plate" to TEM image recording. Ultramicroscopy 25:195
4. Ogura N, Yoshida K, Kojima Y, Saito H (1994) Development of the 25 micron pixel imaging plate system for TEM. In: Proceedings 13th International Congress on Electron Microscopy, Paris, vol 1, p 219
5. Taniyama A, Shindo D, Oikawa T (1997) Detective quantum efficiency of the 25 μm pixel size imaging plate for transmission electron microscopes. J Electron Microsc 46:303

6. Zuo JM, McCartney MR, Spence JCH (1996) Performance of imaging plates for electron recording. Ultramicroscopy 66:35
7. Ishizuka K (1993) Analysis of electron image detection efficiency of slow-scan CCD cameras. Ultramicroscopy 52:7
8. Zuo JM (1996) Electron detection characteristics of slow-scan CCD camera. Ultramicroscopy 66: 21
9. Taniyama A, Oikawa T, Shindo D (1999) Evaluation of the characteristics of a slow-scan CCD camera for a transmission electron microscope. J Electron Microsc 48:257
10. Taniyama A, Shindo D, Oikawa T (1996) Sensitivity and fading characteristics of the 25 μm pixel size imaging plate for transmission electron microscopes. J Electron Microsc 45:232
11. Oikawa T, Shindo D, Hiraga K (1994) Fading characteristic of imaging plate for a transmission electron microscope. J Electron Microsc 43:402
12. Mooney PE, Fan GY, Meyer CE, Truong KV, Bui DB, Krivanek OL (1990) Slow-scan CCD camera for transmission electron microscopy. In: Proceedings 12th International Congress for Electron Microscopy, Seattle, vol 1. San Francisco Press, San Francisco, p 164

3. Electron Energy-Loss Spectroscopy

Electron energy-loss spectroscopy (EELS) is one of the most popular analytical electron microscopy techniques, similar to energy dispersive X-ray spectroscopy (EDS). In the past, EELS was thought to be effective in compositional analysis only for light elements but useless in general for quantitative analysis in comparison with EDS. However, the accuracy of analysis by EELS is much improved recently owing to high performance of the detector and usage of a FEG. Furthermore, an energy-filter system that provides energy-filtered images has been installed on an electron microscope. Thus, currently EELS has attracted much attention for new applications such as elemental mapping and background subtraction in electron diffraction patterns.

In this chapter, inelastic scattering of electrons, which is basic knowledge of EELS, is described first, and then the hardware (spectrometers) and software (techniques) of EELS are noted. In the latter part of this chapter, principles and application of spectrum analysis and energy filtering are explained on the basis of the basic knowledge of EELS.

3.1 Inelastic Scattering of Electrons

As described in section 1.1.1, electrons passing through a specimen are classified into two groups. One is the group of transmitted electrons and elastically scattered electrons that do not suffer any energy loss; the other consists of the electrons scattered inelastically through interaction with a material (Fig. 3.1). With EELS, the spectrum of the electrons is analyzed, paying attention to the inelastically scattered electrons. Among the various energy-loss processes, the typical ones and their energy ranges are as follows.

1. Lattice vibration (phonon excitation): less than 0.1 eV
2. Collective excitation of valence electrons (plasmon excitation): less than 30 eV
3. Interband transition: less than 10 eV
4. Inner-shell electron excitation (core electron excitation): more than 13 eV
5. Excitation of free electron (secondary electron emission): less than 50 eV (background of spectrum)
6. Bremsstrahlung (emission of continuous X-rays): background of spectrum

Figure 3.2 shows a typical energy-loss spectrum obtained from an iron oxide particle. A sharp peak appears (0 eV) in the left-hand side of the spectrum. The energy of this peak, the so-called *zero-loss peak*, corresponds to the incident electron energy. Near the zero-loss peak, there is a peak that results from the plasmon excitation. In a higher energy-loss region, energy-loss peaks appear owing to inner-shell excitation of the constituent elements of oxygen and iron as well as carbon in the microgrid supporting the specimen. Thus, it is seen that the energy-loss processes of items 2 and 4 in the above list provide sharp peaks in a spectrum. From their energy values and intensity distribution, the elements can be identified and their composition determined. Although process 3 (see list, above) does not provide definite peaks, it affects the spectrum in the low enegy-loss region strongly, and its information can be obtained accurately through analysis of the spectrum (see Sect. 3.4). On the other hand, the signal of phonon excitation (item 1) is difficult to detect with a conventional spectrometer that has an energy resolution of about 1 eV. The processes of items 5 and 6 (see list) do not result in sharp peaks in the spectrum, forming only background, so no valuable information can be obtained. In a specific diffraction condition, bremsstrahlung (item 6) may provide some valuable information (see Sect. 4.5.4.1). Taking these situations into account, analytical techniques related to inelastic scattering due to processes noted in items 2–4 in the list are described below. Before explaining measurement and analysis of the spectra, we describe the spectrometers for EELS (see Sect. 3.2), the analytical techniques (see Sect. 3.3), and the theoretical background (see Sect. 3.4).

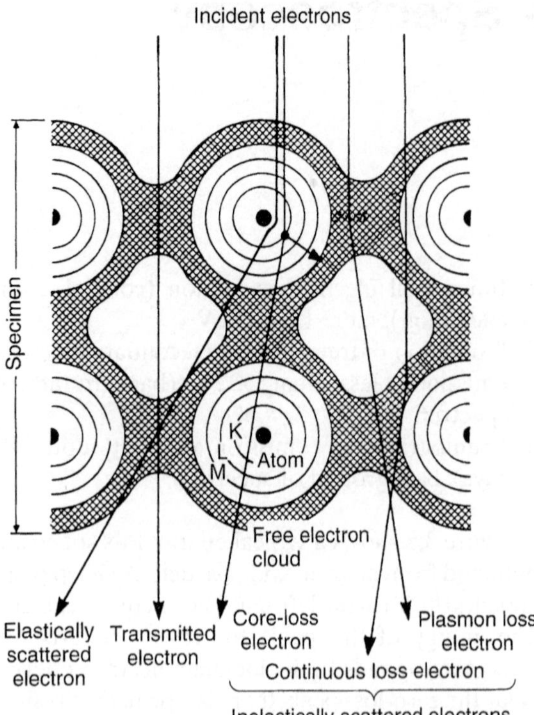

Fig. 3.1. Interaction between incident electrons and a material

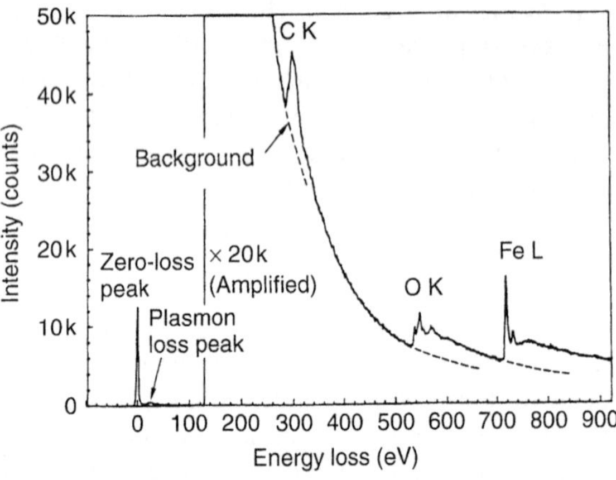

Fig. 3.2. Energy-loss spectrum obtained from an iron oxide particle

3.2 Spectrometer for EELS

3.2.1 Energy Dispersion

When an electron with a velocity v enters a uniform magnetic field whose flux density is B, it suffers Lorentz force in the direction perpendicular to the direction of the electron movement. Thus, the electron starts a round motion in the field. The radius R of the circle is given by

$$R = \frac{\beta_m m_0}{eB} v \tag{3.1}$$

where β_m *is* $\frac{1}{\sqrt{1-\left(\frac{v}{c}\right)^2}}$; m_0 is the rest mass of electron; and c is the velocity of light.

According to Eq. 3.1, the radius R depends only on the velocity v when the magnetic flux density B is constant. If velocities of electrons are different, the radii corresponding to their velocities are different; and so the velocities of the electrons can be analyzed.

On the other hand, the kinetic energy of the electron is given by

$$E = (\beta_m - 1)m_0c^2 \tag{3.2}$$

showing that the kinetic energy is related directly to the velocity. Thus, when electrons with various energies enter the electromagnet of the uniform magnetic field of a spectrometer, energy is dispersed. The dispersion provides a number of electrons as a function of their energy; it is called the *energy spectrum*. The energy of electrons suffering various energy losses in the specimen is analyzed with the spectrometer.

3.2.2 Spectrometer Optics

The spectrometer is sometimes called a "magnetic prism," as its action for energy dispersion is similar to the action of a glass prism for wavelength dispersion. Characteristics of the spectrometer optics are basically different from those of the electron lens in a transmission electron microscope, which is a rotationally symmetrical magnetic lens, although there are some similarities in their lens actions. Figure 3.3 illustrates a spectrometer and the electron trajectory. In Fig. 3.3 the solid lines show the trajectory of zero-loss electrons, and the dashed lines indicate that of energy-loss electrons with a smaller radius. Each trajectory focuses on the exit plane. The angle between the directions of electrons at the entrance and the exit is called the *deflection angle* (Φ). Although various deflection angles can be chosen in principle, a Φ of 90° is usually employed in commercial instruments for design convenience.

The most important points in spectrometer optics are the point source entrance and the point

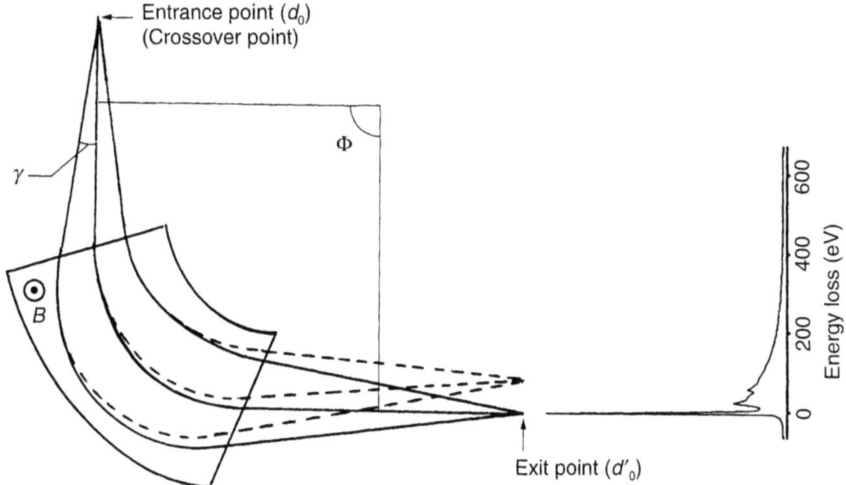

Fig. 3.3. Spectrometer and the electron trajectory

focus exit, similar to that of the magnetic convex lens. The point source, called the entrance point of a spectrometer, is usually set at the crossover point of the projector lens (the final stage of the image-forming lens system). When the electrons have various energies, each focus point forms a plane called the energy dispersion plane. Eventually an energy-loss spectrum is formed on this plane (Fig. 3.3). A larger spot size d_0 at the entrance point produces the correspondingly larger spot size d_0' at the dispersion plane, resulting in lower energy resolution.

An increase in the entrance angle (γ) of the electrons to the spectrometer produces an increase in the second-order aberration of the spectrometer, resulting in lower energy resolution. This is similar to the spherical aberration of the magnetic lens. To correct the second-order aberration, curvatures are introduced for both entrance and exit faces of the spectrometer. For a large entrance angle, however, it is difficult to correct the aberration completely. The distance between the entrance point and the entrance face of the spectrometer and the distance between the exit face of the spectrometer and the energy dispersion plane are completely determined by the shape of a spectrometer, so it cannot be adjusted by changing the excitation of the spectrometer. This situation is similar to that of glass lens optics.

3.2.3 Serial Detection and Parallel Detection

Detecting systems for electron energy-loss spectrum formed on the energy dispersion plane are classified into two groups. One is a serial detection system, and the other is a parallel detection system. Early in the history of spectrometers, photo-film was used to record spectra. During the 1970s a detection system consisting of a single photomultiplier tube (PMT) and a scintillator replaced the use of photo-film. This was necessary because photo-film has low sensitivity; it also had difficulty detecting weak signal intensity and converting it to digital data. Figure 3.4 shows a serial detection system installed on a transmission electron microscope. The spectrometer is set at the final image plane of the electron microscope (i.e., behind the camera chamber). The PMT is a one-dimensional detector, and an energy-selecting slit is located at the energy-dispersion plane in front of the PMT. By changing the exitation current of a spectrometer gradually, the electrons with different energies enter the slit continuously. Thus electron intensity detected by the PMT is displayed by synchronizing it with the energy scanned. In this way the electron intensity distribution as a function of energy loss (i.e., the energy-loss spectrum) is obtained. In this detection system, the energy axis is changed with time series, and so is called *serial detection*. The system had the disadvantage of lower sensitivity because of the time sequence detection mode. To overcome this problem, during the middle of the 1980s a parallel detection system [1,2] was developed using a parallel detector of semiconductor, as the semiconductor detector had been put to practical use at this time.

Figure 3.5 shows a parallel detection system. The parallel detection system consists of a scintil-

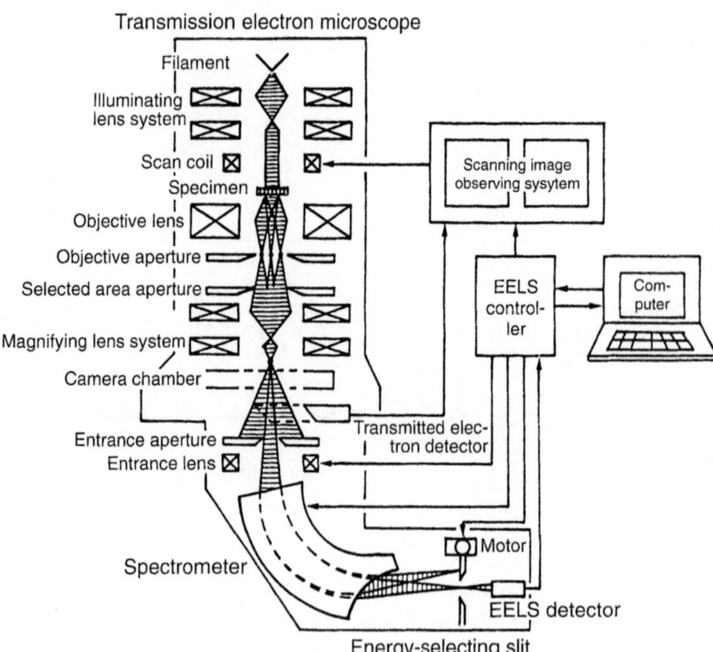

Fig. 3.4. Serial detection system installed on a transmission electron microscope. *EELS*, electron energy-loss spectroscopy

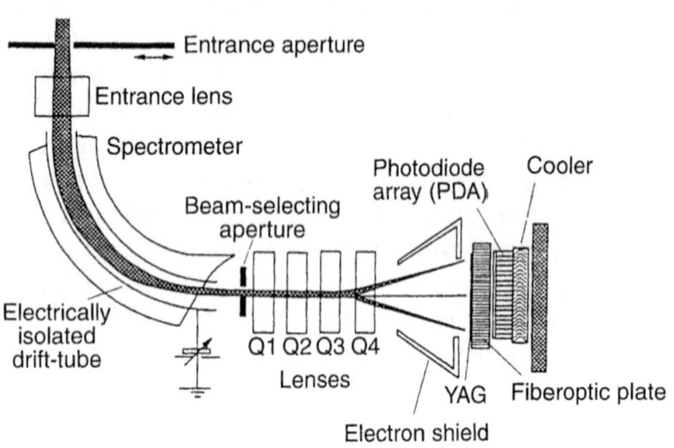

Fig. 3.5. Parallel detection system and an electron ray diagram

lator (YAG) and photodiode array detector (1024 or 2048 channels) connected to a fiberoptic plate. The signals in all the channels are detected simultaneously, so in principle detection is more efficient than serial detection channel by channel. Although parallel detection has the advantage of high detection efficiency, it has the disadvantage of a lower dynamic range in the semiconductor detector for electron intensity than did the serial detection system. Therefore, it is sometimes difficult to record a zero-loss peak of high intensity and a plasmon loss peak or core loss peak of low intensity (or both) in the same spectrum simultaneously. When the signal intensity is low, a spectrum detected is influenced by diode characteristics, such as sensitivity deviation from channel to channel, dark current, and readout noise. These characteristics should be corrected in the spectrum before the analysis. Moreover, CCD is employed as a parallel detector instead of the photodiode array.

Energy resolution of the spectrometer is usually about 1–2eV for 200kV transmission electron microscopes. Figure 3.6 shows a zero-loss spectrum with 0.7eV resolution at full width at half-maximum (FWHM), obtained with a 200kV electron microscope with an FEG. The spectrum resolution is determined by the convolution of the spectrometer resolution and the energy spread of the electrons. The incident beam must be high

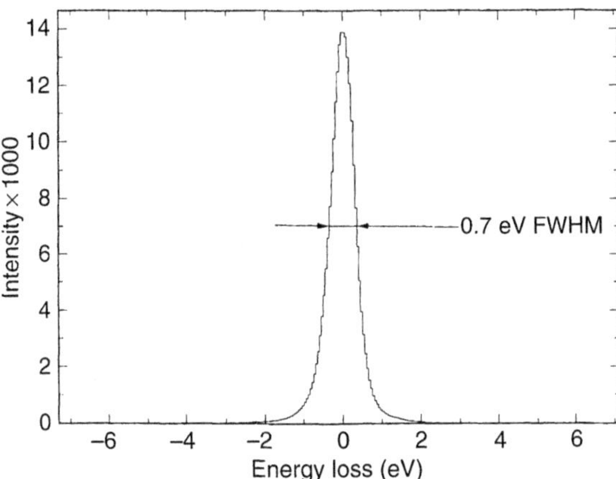

Fig. 3.6. Zero-loss spectrum obtained with a 200 kV electron microscope with a thermal FEG

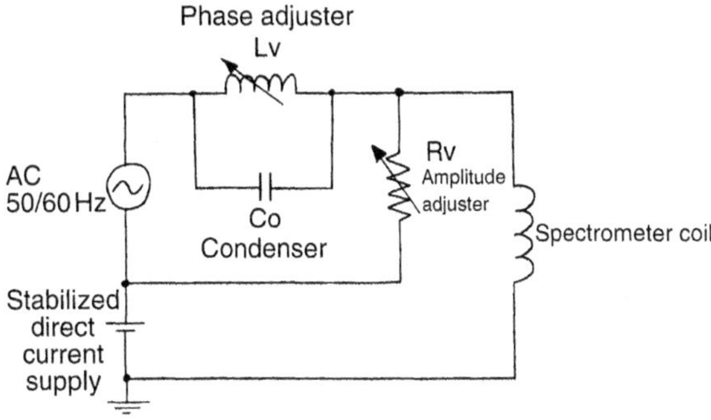

Fig. 3.7. Example of a generator circuit of the canceling current

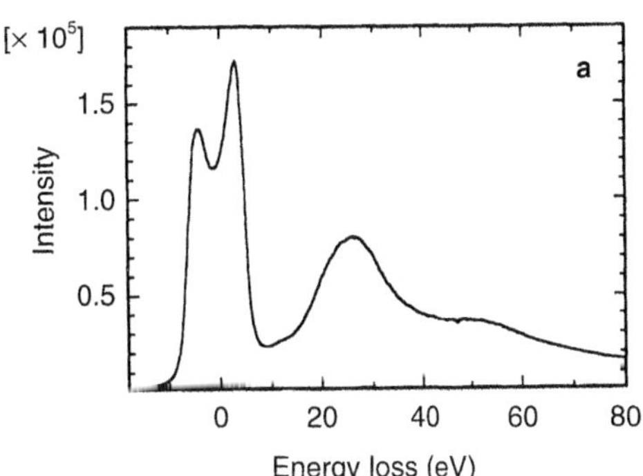

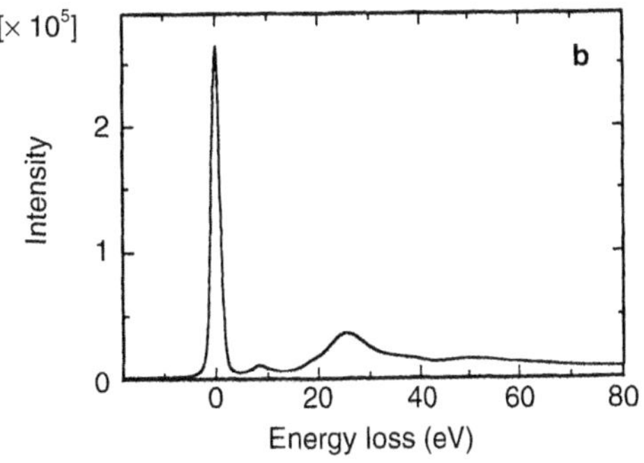

Fig. 3.8. **a** Spectrum affected by the external field. **b** Spectrum corrected by noise canceling

enough to observe a core-loss signal in a high energy-loss region of the spectrum; thus, the emission current must be high and the beam highly converged. In this case, the energy spread of the electrons increases, so the energy resolution becomes 0.8–1.0 eV even if an FEG is used. To obtain higher energy resolution, an analytical electron microscope combined with a monochrometer has been developed, and high energy resolution of less than 0.1 eV at FWHM has been obtained [3].

3.2.4 Compensation of External Magnetic Field

The resolution of the spectrometer attainable is about 1 eV for a 200 kV transmission electron microscope. Thus, the ratio of the resolution to the accelerating voltage δ_R is

$$\delta_R = \frac{1}{200{,}000} = 5 \cdot 10^{-6} = 5\ \text{ppm} \tag{3.3}$$

This indicates that high stability is required in the magnetic field of the spectrometer. If there is an external magnetic field or an AC stray field from a power supply circuit (or both), the stability of the magnetic field in the spectromenter is disturbed and its energy resolution decreases. Although the spectrometer is enveloped in a magnetic shielding case, the magnetic noise canceling circuit is necessary for attaining high energy resolution. A canceling current to the spectrometer excitation current is usually added for this purpose. Figure 3.7 shows a generator circuit of the canceling current. In this circuit the frequency is set to be the same as the electric power (50 or 60 Hz). The phase and amplitude are adjusted so as to cancel the disturbance. Figure 3.8 shows the effects of noise canceling. When there is a large amount of magnetic noise, the zero-loss peak splits, and fine structures of core loss peaks disappear.

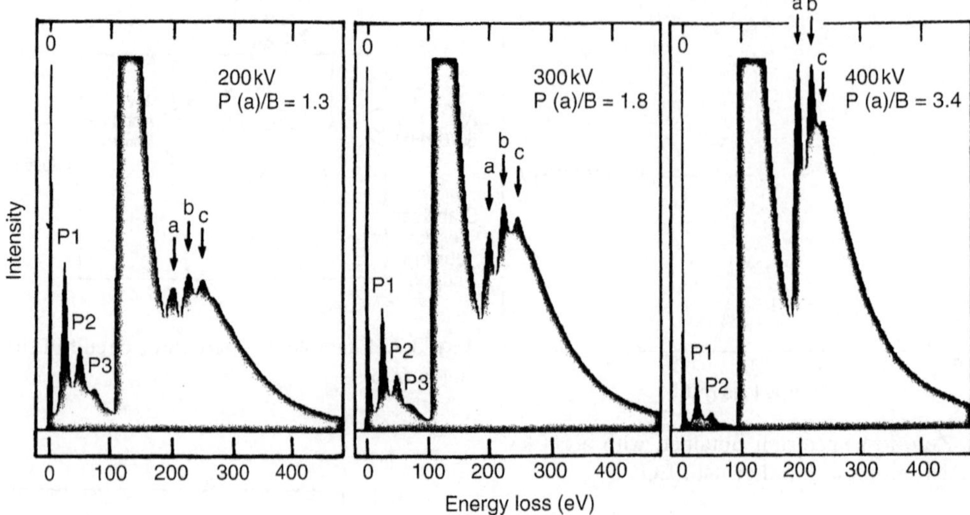

Fig. 3.9. Dependence of boron K-edges on accelerating voltage obtained from a boron single crystal 0.5 μm in thickness

3.3 Analytical Technique in EELS

3.3.1 Accelerating Voltage

With the increased accelerating voltage in a transmission electron microscope, the energy resolution of a spectrometer is lowered. The peak/background ratio (P/B), which is also called the jump ratio, increases because the probability of the multiple electron scattering decreases and the effective acceptance angle of the spectrometer increases.

Figure 3.9 shows the dependence of boron K-edges on the accelerating voltage obtained from a boron single crystal. The spectra obtained at 200, 300, and 400 kV are normalized with the zero-loss intensity. The data show that the P/B ratio in peaks (a–c) of a boron K-edge increases with the increase in accelerating voltage. Because the P/B ratio increases with the accelerating voltage, it is expected that weak intensity of a core-loss peak in a high energy-loss region, which is difficult to observe with a low accelerating voltage, can be detected. Figure 3.10 shows an energy-loss spectrum obtained from iron at an accelerating voltage of 400 kV. In the spectrum, in addition to Fe M- and L-edges and an oxygen K-edge from surface oxide film on the specimen that can be observed with a 200 kV microscope, an Fe K-edge at much higher energy 7114 eV is clearly observed. In the plasmon peaks of Fig. 3.9 (peaks of P2 and P3 result from double and triple plasmon excitations, respectively) (see Sect. 3.5.1) near the zero-loss peak, intensities of the peaks decrease with the increase in accelerating voltage. This indicates that the mean free path for inelastic scattering due to plasmon excitation increases with the increase in accelerating voltage. Figure 3.11 shows the mean free paths for inelastic electron scattering (see Sect. 3.5.2) in Si and SiO_2 observed as a function of accelerating voltage [4]. The mean free paths increase with increased accelerating voltage in both cases.

3.3.2 Acceptance Angle

Figure 3.12 shows the mean free path for inelastic scattering in aluminum as functions of the accelerating voltage and the acceptance angle. The filled circles are experimental data, and the solid lines show theoretical values [5], given by

$$\lambda_p = \frac{a_0}{\Theta_E \ln\left(\dfrac{\beta}{\Theta_E}\right)} \tag{3.4}$$

$$\Theta_E = \frac{E_K(1+2\varepsilon E_0)}{2E_0(1+\varepsilon E_0)} \tag{3.5}$$

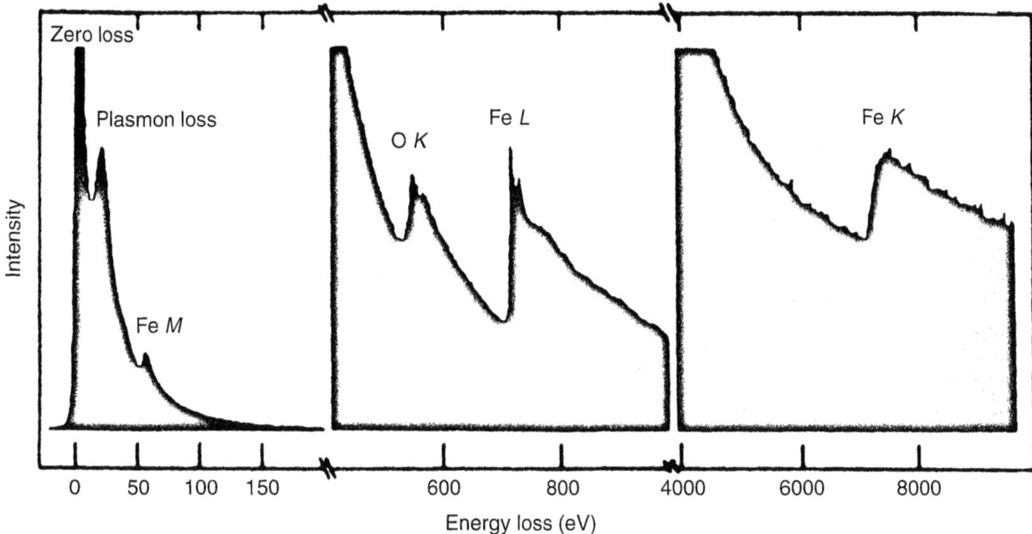

Fig. 3.10. Energy-loss spectrum obtained from iron at an accelerating voltage of 400 kV

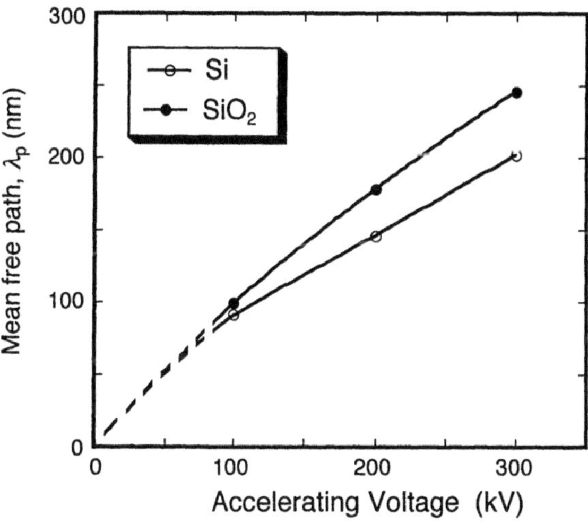

Fig. 3.11. Mean free paths of inelastic electron scattering in Si and SiO_2 observed as a function of accelerating voltage. No objective aperture is used

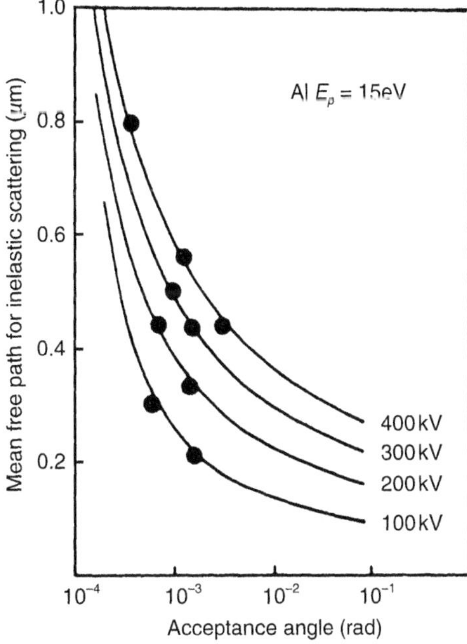

Fig. 3.12. Mean free path for inelastic scattering in aluminum as functions of accelerating voltage and acceptance angle

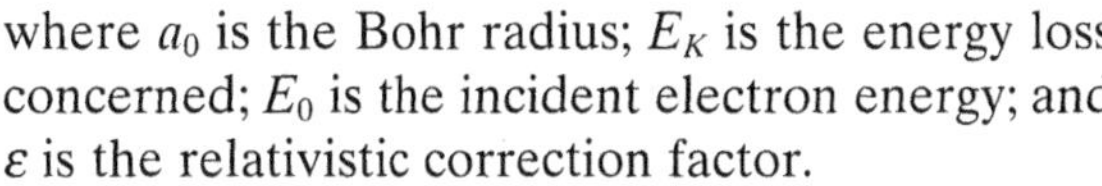

where a_0 is the Bohr radius; E_K is the energy loss concerned; E_0 is the incident electron energy; and ε is the relativistic correction factor.

When the acceptance angle is fixed, the mean free path for inelastic scattering increases with increasing accelerating voltage, as for Si and SiO_2 in Fig. 3.11. On the other hand, it is seen that the mean free path for inelastic scattering decreases with the increase in acceptance angle for each accelerating voltage. This suggests that the distribution of the inelastically scattered electrons extends outside the central beam (see details in Sect. 3.6.3.2).

In elemental or state analysis, the signal intensity detected I_s is generally given by

$$I_s = n \cdot t \cdot \sigma(\beta, \Delta E, E_0) \cdot I_T \qquad (3.6)$$

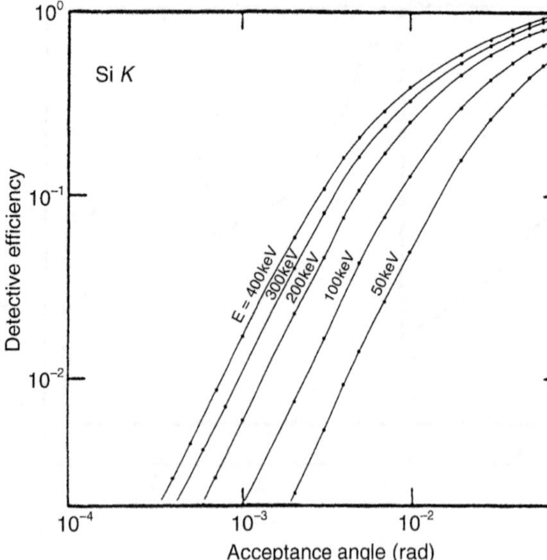

Fig. 3.13. Detection efficiency (η_β) of silicon K-edge calculated as a function of acceptance angle β

where ΔE is the energy window width; n and t are the number of atoms in a unit volume of the element and specimen thickness, respectively; and $\sigma(\beta, \Delta E, E_0)$ and I_T are the partial cross section and incident electron intensity, respectively [6]. The partial cross section is given with the total cross section $\sigma_t(\pi, \infty, E_0)$ as

$$\sigma(\beta, \Delta E, E_0) = \eta_\beta \cdot \eta_{\Delta E} \cdot \sigma_t(\pi, \infty, E_0) \quad (3.7)$$

where η_β and $\eta_{\Delta E}$ are detective efficiencies for the acceptance angle and the width of energy window, respectively. Figure 3.13 shows the detection efficiency (η_β) of a silicon K-edge calculated as a function of acceptance angle β [5]. It is seen that the efficiency increases with an increase in acceptance angle or incident electron energy.

3.3.3 Analytical Modes

The image-forming lens system of a transmission electron microscope works as the prelens system of the spectrometer. It has two capabilities: forming images and supplying the incident beam for the spectrometer. Therefore, selecting a mode (imaging mode or diffraction mode) is important in the analysis. In the *imaging mode*, an image is formed on the screen of a transmission electron microscope, and a diffraction pattern is formed at the entrance point of the spectrometer. The field-of-view for the analysis can be selected with an entrance aperture of the spectrometer. The size of the area for the analysis is adjustable by changing the magnification of a microscope in this mode. In the *diffraction mode*, a diffraction pattern is formed on the screen, and an image is formed at the entrance point. A scattering angle (acceptance angle) can be limited with the entrance aperture. An objective aperture should be used to limit a scattering angle in the imaging mode.

3.3.4 Detection Limit in EELS

The detection limit has been discussed with the minimum detectable mass (MDM) (see ref. [7], for example). It has been determined by the scattering cross section, which depends on the atomic number (Z), specimen thickness, and acquisition time. Recently, the performance of electron microscopes has been improved by utilizing an FEG and a parallel detector of high sensitivity. Suenaga et al. [8] have reported an example of atomic order detection. In their study, a single Gd atom in a single wall carbon nanotube was detected by a 100 kV scanning transmission electron microscope with an FEG. A minimum detectable concentration in a standard specimen by EELS is considered to be, in rough estimation, a few atomic percentage (at%) points.

3.4 Theoretical Background of EELS

The differential cross section for the inelastically scattered electrons with energy loss is generally given with the dielectric function ε ($\varepsilon = \varepsilon_1 + i\varepsilon_2$) as

$$\frac{\partial^2 \sigma}{\partial E \partial \Omega} = \frac{1}{\pi^2 a_0^2 c^2} \cdot \frac{1}{q^2} \cdot \mathrm{Im}\left[-\frac{1}{\varepsilon}\right] \quad (3.8)$$

where $\mathrm{Im}[-1/\varepsilon]$ indicates the imaginary part of $-1/\varepsilon$, called the energy-loss function [9] or the loss function [10]; a_0 is the Bohr radius (see Appendix 1); and q is the scattering vector shown in Fig. 3.27 (see later). The energy-loss function can be given with the real part ε_1 and the imaginary part ε_2 as

$$\mathrm{Im}\left[-\frac{1}{\varepsilon}\right] = \frac{\varepsilon_2}{|\varepsilon|^2} = \frac{\varepsilon_2}{\varepsilon_1^2 + \varepsilon_2^2} \quad (3.9)$$

The dielectric function is given with the electric displacement D_ω of the incident electron due to the charge density and the resultant electric field E_ω induced as

$$\varepsilon = \frac{D_\omega}{E_\omega} \tag{3.10}$$

Consider the case where the free electron is excited by the incident electron (Drude model). When the electric field of the oscillation with the frequency ω for the free electron is given by $E_\omega = E_0\exp(-i\omega t)$ and their relaxation time is τ, the displacement of the free electron satisfies the following equation.

$$m_0\left(\frac{d^2x}{dt^2} + \frac{1}{\tau}\frac{dx}{dt}\right) = -eE_\omega \tag{3.11}$$

The solution of this equation is given by

$$x = \frac{eE_\omega}{m_0\left(\omega^2 + i\frac{\omega}{t}\right)} \tag{3.12}$$

The dipole moment of the electrons P induced by this oscillation is given as

$$\begin{aligned} P &= -nex \\ &= \frac{ne^2}{m_0}E_\omega \cdot \frac{1}{\omega^2 + i\frac{\omega}{\tau}} \end{aligned} \tag{3.13}$$

where n is the electron density. Thus, the dielectric function for the free electron is obtained as

$$\begin{aligned} \varepsilon(\omega) &= \frac{D_\omega}{E_\omega} = 1 + 4\pi\frac{P}{E_\omega} \\ &= 1 - \frac{4\pi ne^2}{m_0} \cdot \frac{1}{\omega^2 + i\frac{\omega}{\tau}} \\ &= 1 - \omega_p^2 \cdot \frac{1}{\omega^2 + i\frac{\omega}{\tau}} \end{aligned} \tag{3.14}$$

where ω_p is called the plasmon frequency and is given by

$$\omega_p^2 = \frac{4\pi ne^2}{m_0} \tag{3.15}$$

Using the relation of $E = \hbar\omega$, ε can also be expressed with the energy loss E as

$$\varepsilon(E) = 1 - \frac{E_p^2}{E^2 + iE\frac{\hbar}{\tau}} \tag{3.16}$$

In the above equation,

$$E_p = \hbar\omega_p \tag{3.17}$$

where E_p is the so-called *plasmon energy*, which corresponds to the energy obtained by quantizing the collective motion of the electrons inside the specimen. Specifically, it is called the *volume plasmon*, and the quantized collected motion of the electrons at the surface is called the *surface plasmon*. From the above equation, the real part ε_1 and the imaginary part ε_2 of the dielectric function are given by

$$\varepsilon_1(E) = 1 - \frac{E_p^2}{E^2} \cdot \frac{1}{1 + \left(\frac{\Delta E_p}{E}\right)^2} \tag{3.18}$$

$$\varepsilon_2(E) = \frac{\Delta E_p}{E} \cdot \frac{E_p^2}{E^2} \cdot \frac{1}{1 + \left(\frac{\Delta E_p}{E}\right)^2} \tag{3.19}$$

where ΔE_p (= $\hbar/\tau$) corresponds to the FWHM of the energy-loss function. When the relaxation time for damping the plasma oscillation is long enough (i.e., ΔE_p is very small), ε_1 is zero at $E = E_p$. Then, from Eq. 3.9

$$\mathrm{Im}\left[-\frac{1}{\varepsilon(E)}\right] \simeq \frac{1}{\varepsilon_2} \tag{3.20}$$

Also, from Eq. 3.19, $1/\varepsilon_2$ takes on a large value, and eventually the energy-loss function forms a sharp peak similar to the δ-function at the plasma energy. Strictly speaking, the relaxation time is finite, and thus ε_1 becomes zero for

$$E(\varepsilon_1 = 0) = \left[(E_p)^2 - (\Delta E_p)^2\right]^{1/2} \tag{3.21}$$

In Fig. 3.14 ε_1, ε_2, and the energy-loss function are shown for E_p = 15 eV and ΔE_p = 1 eV. Note that the peak energy of the energy-loss function is nearly equal to E_p = 15 eV, and ε_1 is zero at this energy. On the other hand, at a higher energy region, ε_1 is almost equal to 1, and ε_2 takes on a small value. The energy-loss function calculated for ΔE_p = 4 eV is shown in Fig. 3.14 with a dotted line. It is seen that the FWHM becomes larger and the peak position shifts to a lower energy region

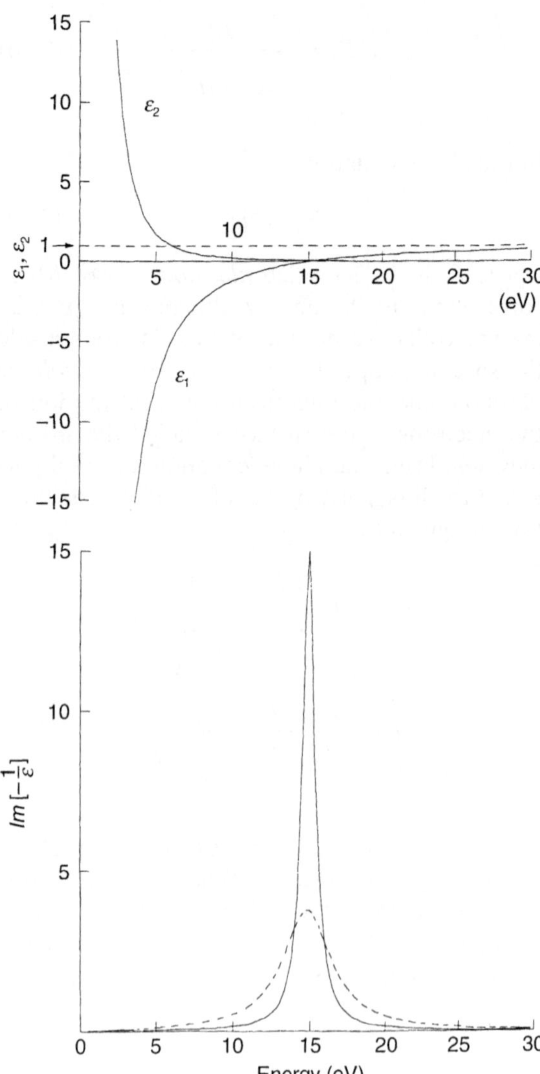

Fig. 3.14. Values of ε_1, ε_2 and the energy-loss function calculated for E_p = 15 eV and ΔE_p = 1 eV. Energy-loss function (*dotted line*) is for ΔE_p = 4 eV

(compare the heights of the cross points of the solid and dotted lines).

When the interband transition of the valence electron is taken into account, the oscillation term corresponding to the excitation energy $E_i = \hbar\omega_i$ is added to Eq. 3.11. Thus, the equation for this case is

$$m_0\left(\frac{d^2x}{dt^2}+\frac{1}{\tau}\frac{dx}{dt}+\omega_i^2 x\right)=-eE_\omega \tag{3.22}$$

The ε_1 and ε_2 obtained from this equation are

$$\varepsilon_1(E)=1-\left(\frac{E_p}{E}\right)^2 \cdot \frac{1-\left(\frac{E_i}{E}\right)^2}{\left[1-\left(\frac{E_i}{E}\right)^2\right]^2+\left(\frac{\Delta E_p}{E}\right)^2} \tag{3.23}$$

$$\varepsilon_2(E)=\frac{\Delta E_p}{E}\cdot\left(\frac{E_p}{E}\right)^2 \cdot \frac{1}{\left[1-\left(\frac{E_i}{E}\right)^2\right]^2+\left(\frac{\Delta E_p}{E}\right)^2} \tag{3.24}$$

Assuming that the valence electron density corresponds to the plasmon energy used in Fig. 3.14, ε_1 and ε_2 and the energy-loss function are shown with E_i = 5 eV in Fig. 3.15. At a higher energy region, ε_1 is nearly equal to 1 and ε_2 takes on a small value. On the other hand, at a lower energy region, ε_2 forms the peak at the energy of the interband transition E_i. The plasma energy shifts to higher energy than Fig. 3.14. Moreover, ε_1 becomes zero at energy E_i and around energy

$$E=\left(E_p^2+E_i^2\right)^{1/2} \tag{3.25}$$

which corresponds to the peak position of the energy-loss function. In this way, the interband transition does not produce the peak in the energy-loss spectrum, but it is clarified from ε_2. Actually, ε_2 corresponds to the optical absorption spectrum, and gives direct information about the interband transition. On the other hand, in the energy-loss spectrum ε_2 is not directly obtained, but the value multiplied by $1/(\varepsilon_2^2+\varepsilon_1^2)$ is obtained. Thus, to evaluate ε_2 from the energy-loss spectrum in the lower energy region, the real part of $1/\varepsilon$ should first be obtained with the so-called Kramers-Kronig transformation. To carry out such analysis, the energy-loss spectrum in the lower energy region should be measured with high precision.

The excitation energy for the surface plasmon, which is produced by inducing the longitudinal wave of collection motion of electrons along the metal surface, is given with the volume plasmon energy E_p as

$$E_S=E_p/\sqrt{2} \tag{3.26}$$

Thus, the surface plasmon peak appears at a lower energy region than the volume plasmon.

For inner-shell excitation, simple treatment with single electron excitation is possible. In the high-

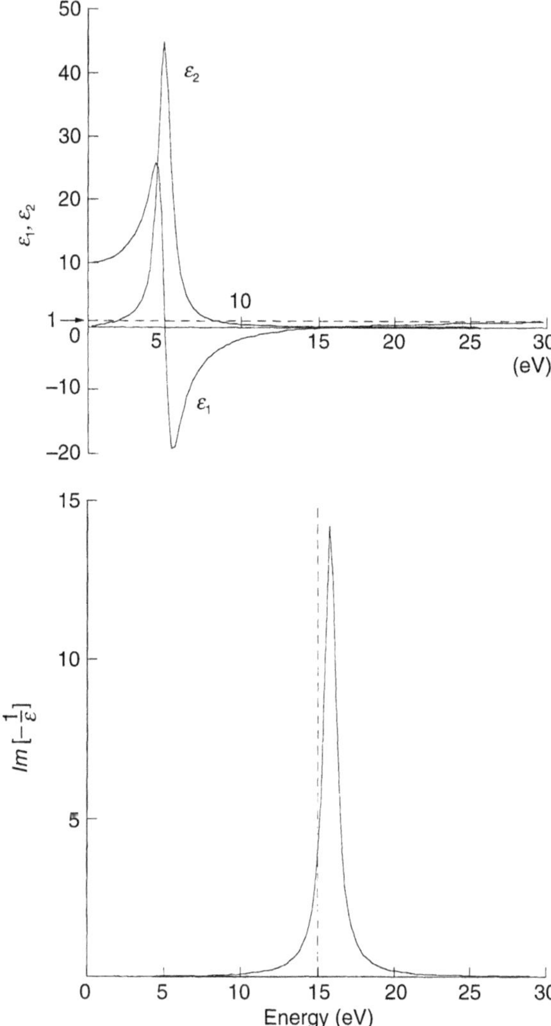

Fig. 3.15. Values of ε_1, ε_2 and the energy-loss function calculated taking into account the interband transition ($E_i = 5\,\text{eV}$)

energy region where the inner-shell excitations appear, ε_1 of the dielectric function is nearly equal to 1, and ε_2 is much smaller than 1. Thus, based on Eq. 3.9, the energy-loss function can be interpreted with ε_2, which corresponds to the optical absorption spectrum. In other words, the inelastic cross section for inner-shell excitation is given with the transition probability in quantum mechanics as

$$\frac{d^2\sigma}{dEd\Omega} = \frac{1}{\pi^2 a_0^2 e^2} \cdot \frac{1}{q^2} \cdot \varepsilon_2 = \frac{4}{a_0^2 q^4} \cdot \sum_{if} |<f|\exp(iqr)|i>|^2 \delta(E - E_f + E_i) \quad (3.27)$$

where $|i>$ and $|f>$ are the wave functions of the initial and final states for the single electron excitation process. Furthermore, the energy of the initial state is localized and the exp term can be expanded for a small scattering angle (dipole approximation), and thus Eq. 3.27 can be simplified as

$$\frac{d^2\sigma}{dEd\Omega} = \frac{4}{a_0^2 q^4} \sum_f |<f|qr|i>|^2 \delta(E - E_f + E_i) \quad (3.28)$$

The orthogonalities of the initial state and final state are utilized in the derivation of Eq. 3.28. From Eq. 3.28 it is seen that the transition of the core electron into the unoccupied states above the Fermi level with the selection rule satisfying the relation $\Delta l = \pm 1$ (where l is the quantum number of orbital angular momentum) can be observed in the energy-loss spectrum. For example, when the s electron in the inner-shell is excited, the unoccupied density of states in the p orbital can be observed; and for excitation of the p electron, the unoccupied density of the s and d states can be observed in the energy-loss spectra. The correspondence of the excitation of the 1s electron and the K-edge observed in the energy-loss spectrum are shown in Fig. 3.16. It is noted that because of the limited resolution of EELS, the detailed fine structure of unoccupied density of states does not appear in the spectrum. Also, in the metallic specimens, the core hole produced by inner-shell excitation is thought to be screened by the other atomic electrons (screening effect); but in some other cases the screening effect is rather weak, and the interaction between the core hole and the electron excited cannot be neglected. Thus, care should be taken when interpretating the energy-loss spectrum. Furthermore, for large scattering angles a transition that does not satisfy the selection rule sometimes appears [11].

3.5 Analysis of Electron Energy-Loss Spectra

3.5.1 Energy Loss Due to Plasmon Excitation

The main excitation of valence electrons is plasmon excitation, which is observable in a low energy-loss region. As described in Section 3.4, a peak corresponding to plasmon excitation appears near the energy of plasmon energy E_p (see Eq. 3.19). Therefore, it is possible to identify materials and obtain information of a composi-

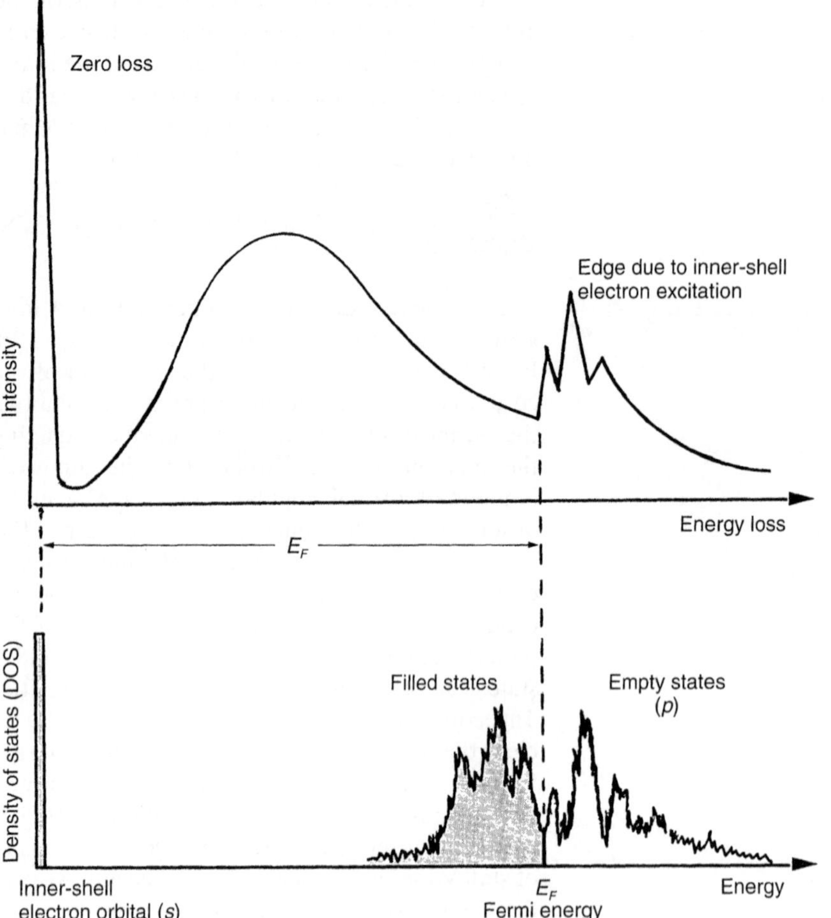

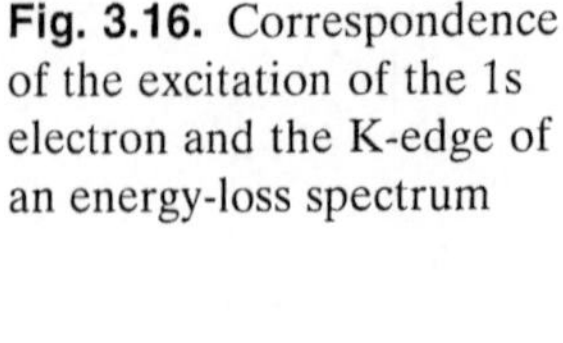
Fig. 3.16. Correspondence of the excitation of the 1s electron and the K-edge of an energy-loss spectrum

tional change in a specimen from the plasmon excitation energies. Figure 3.17 shows an energy-loss spectrum with valence electron excitation from a small sodium (Na) metal crystal[1] produced by electron irradiation.

The energy value (5.70 eV) for the energy-loss peak due to valence electron excitation near the zero-loss peak is slightly smaller than a theoretical value for plasmon excitation in sodium (5.95 eV) (Table 3.1), as noted in Section 3.4. A peak appearing at about 11.4 eV is the so-called

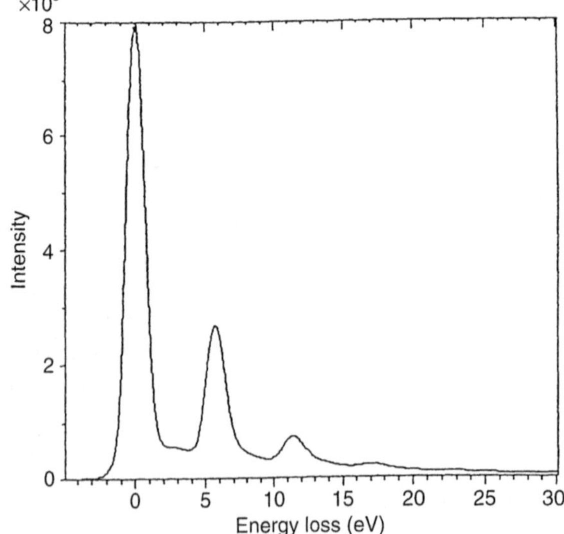

Fig. 3.17. Energy-loss spectrum of a small sodium metal crystal

[1] It is generally impossible to insert and observe sodium metal in a transmission electron microscope because sodium is easily oxidized in air even at room temperature. However, if an $NaAlH_4$ compound is irradiated with electrons in a transmission electron microscope, sodium single crystals precipitate from the compound and the sodium crystal can be observed by electron microscopy [12].

second plasmon peak. It appears at energy two times higher than the energy of the first (original) plasmon peak, as the peak originates from double excitation of the plasmon by the incident electrons in a specimen. For thick specimens, such peaks resulting from the plural scattering are observed frequently. Figure 3.18 shows energy-loss spectra with valence electron excitation in an aluminum crystal. With increased specimen thickness, the number of the peaks due to plural plasmon excitations increases. Theoretical and experimental values of energy-loss peaks for valence electron excitation are compared for some typical materials in Table 3.1 [13–18]. Excitation of valence electrons has a scattering cross section larger than the inner-shell excitation observable at a higher energy-loss region, so it is generally easy to observe the plasmon loss spectrum.

3.5.2 Measurement of Mean Free Path of Inelastic Scattering and Estimation of Specimen Thickness

As shown in Fig. 3.18, the intensity of the plasmon peaks increases with an increase in specimen thickness. On the other hand, the zero-loss intensity decreases were complementary to the plasmon peak intensity. In general, the intensity of the zero-loss peak I_0 for a specimen of thickness t is given with the total electron intensity I_T as

$$I_0 = I_T \exp\left(-\frac{t}{\lambda_p(\beta)}\right) \tag{3.29}$$

where $\lambda_p(\beta)$ is the constant being called the mean free path for inelastic scattering, and it depends on the acceptance angle β. In general, plasmon exci-

Table 3.1. Energy-loss value and FWHM (ΔE_p) for valence electron excitation.

Material	Experimental		Theoretical	Reference
	E_p (eV)	ΔE_p (eV)	E_p (eV)	
Na	5.72	0.4	5.95	13
Al	14.95 ± 0.05	0.5 ± 0.1	15.8	14
Diamond (C)	34	14	31	15
Si Crystal	16.45 ± 0.1	3.6	16.6	16
Si Amorphous	16.1 ± 0.1	4.0		16
Ge Crystal	15.9 ± 0.1	3.4 ± 0.2	15.6	17
Ge Amorphous	15.8 ± 0.2	4.1 ± 0.8	14.8	18

E_p, energy loss; ΔE_p, full width at half-maximum (FWHM)

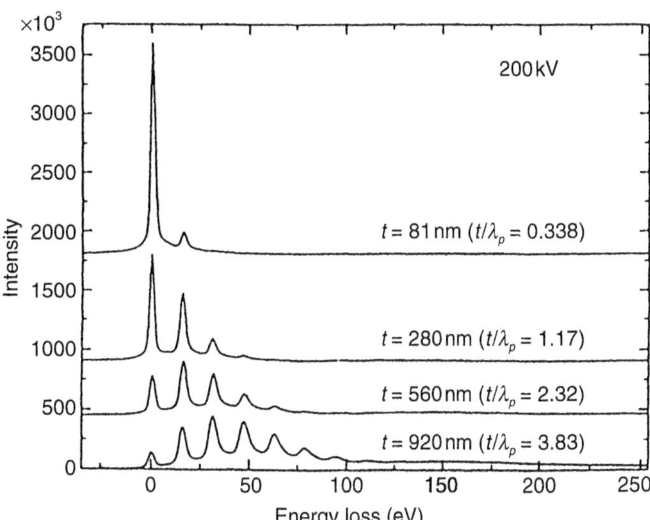

Fig. 3.18. Energy-loss spectra of aluminum crystals with various crystal thicknesses

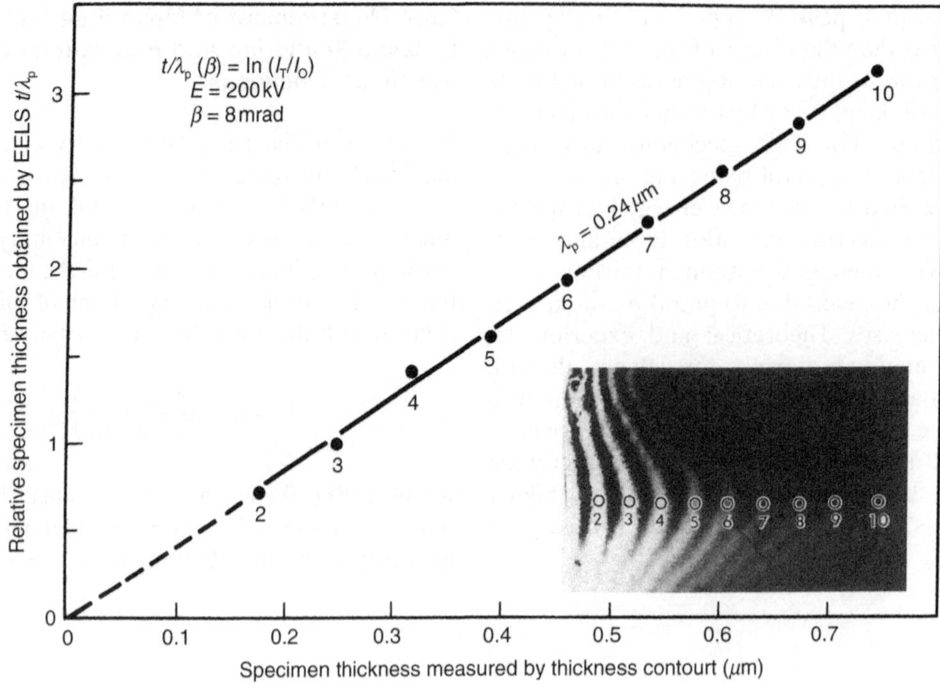

Fig. 3.19. Relation between relative specimen thickness t/λ_p obtained with EELS and specimen thickness t obtained with extinction contour in aluminum

tation is the most probable inelastic scattering process in the specimen, and thus the mean free path for plasmon excitation is the main component in $\lambda_p(\beta)$. From Eq. 3.29, specimen thickness t is given by

$$t = \lambda_p(\beta)\ln(I_T/I_0) \tag{3.30}$$

In the above equation, I_T and I_0 can easily be evaluated from the spectrum, and so the specimen thickness can be determined with high accuracy if λ_p is known.

Figure 3.19 shows the relation between so-called relative specimen thickness t/λ_p and specimen thickness t obtained with aluminum. It is seen that t/λ_p or $\ln(I_T/I_0)$ is proportional to t, which is measured from the extinction contour obtained under the two-beam diffraction condition with the 111 reflection excited. Taking into account an extinction distance for the 111 reflection of aluminum, the mean free path is obtained as $\lambda_p(\beta = 8\,\text{mrad}) = 240\,\text{nm}$ for an accelerating voltage of 200 kV [19].

Figure 3.20 shows a relation between $\ln(I_T/I_0)$ and specimen thickness t obtained from an α-Fe_2O_3 particle of platelet shape whose thickness is known. The shape of the specimen is shown in

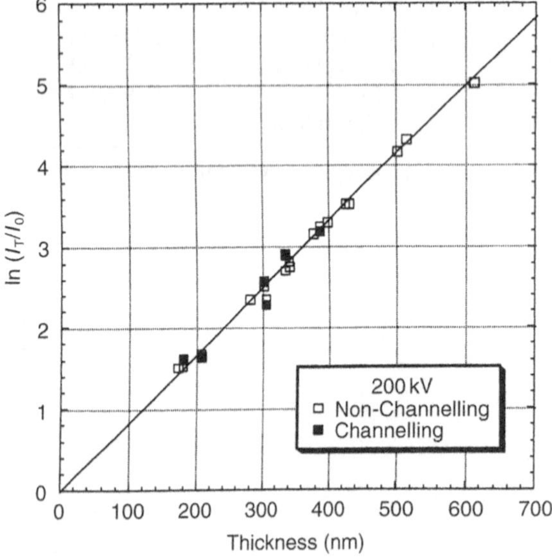

Fig. 3.20. Relation between $\ln(I_T/I_0)$ and specimen thickness t obtained from an α-Fe_2O_3 particle

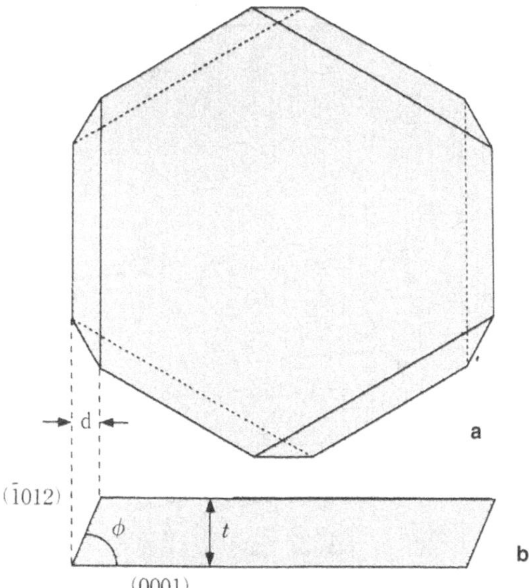

Fig. 3.21. Geometric morphology of a platelet-type α-Fe_2O_3 particle. **a** External shape. **b** Cross section ($\phi = 57.5°$)

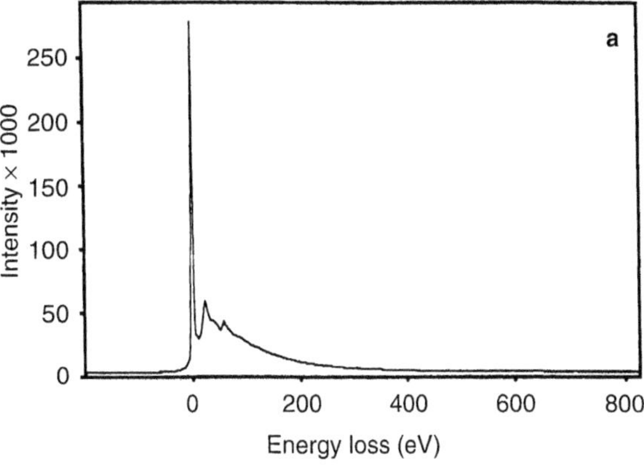

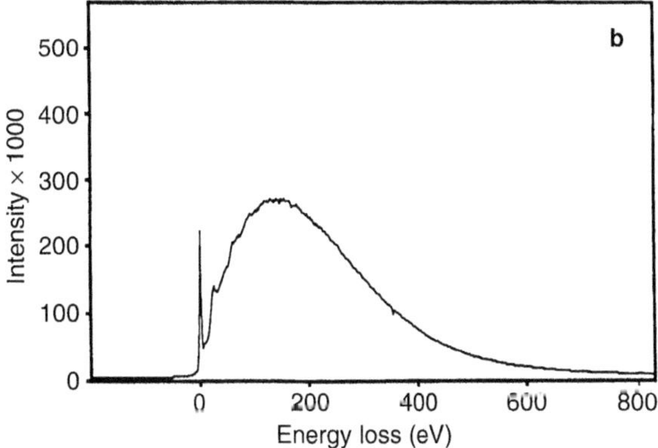

Fig. 3.22. Energy-loss spectra obtained from α-Fe_2O_3. **a** Thin region. **b** Thick region

Fig. 3.21a. When the specimen is observed with the incident electron beam parallel to the [0001] direction (see Fig. 3.21b), the thickness t is estimated from the width of the side surface of the particle projected along the [0001] direction. Figure 3.22 shows two energy-loss spectra obtained from specimens with different thicknesses. In Fig. 3.20, there is a good proportional relation between t and $\ln(I_T/I_0)$, and this relation holds even for the channeling condition with the incidence beam parallel to the zone axis. Based on this relation, the mean free path of α-Fe_2O_3 for an accelerating voltage of 200 kV is evaluated to be $l_p = 120 \pm 10$ nm [20]. Using this λ_p value, estimating the thickness is possible even for particles with complicated shapes and for thin film sliced by ultramicrotomy. The value of λ_p noted above is measured without an objective aperture. Figure 3.23 shows $1/\ln(I_T/I_0)$ or λ_p/t for α-Fe_2O_3 as a function of the acceptance angle β. It is seen that λ_p/t decreases with the increase in β. This is the same tendency as for Al shown in Fig. 3.12. However, in the case of axial illumination (Fig. 3.23b), λ_p/t does not decrease monotonously with an increase in β for especially small acceptance angles. This phenomenon is thought to result from the dynamical diffraction effect.

The specimen thickness can be measured by other methods such as convergent beam electron diffraction (CBED) (see Sect. 5.1.2). However, the usage of EELS for thickness measurements seems to be more effective if λ_p is obtained, as application of the method is easy and its accuracy is not directly affected by the crystallinity or the existence of lattice defects.

3.5.3 Energy Loss Due to Inner-Shell Electron Excitation

3.5.3.1 Elemental Analysis

As shown in Fig. 3.16, the threshold energy of inner-shell excitation corresponds to the difference between the energy of the inner shell and the Fermi energy; thus, elements can be identified by measuring this threshold energy. Figure 3.24 shows the qualitative analysis of a ceramic called SIALON. The signals of nitrogen and oxygen are detected from area G (inside the grain) and area B (grain boundary) 30 nm in diameter. Different

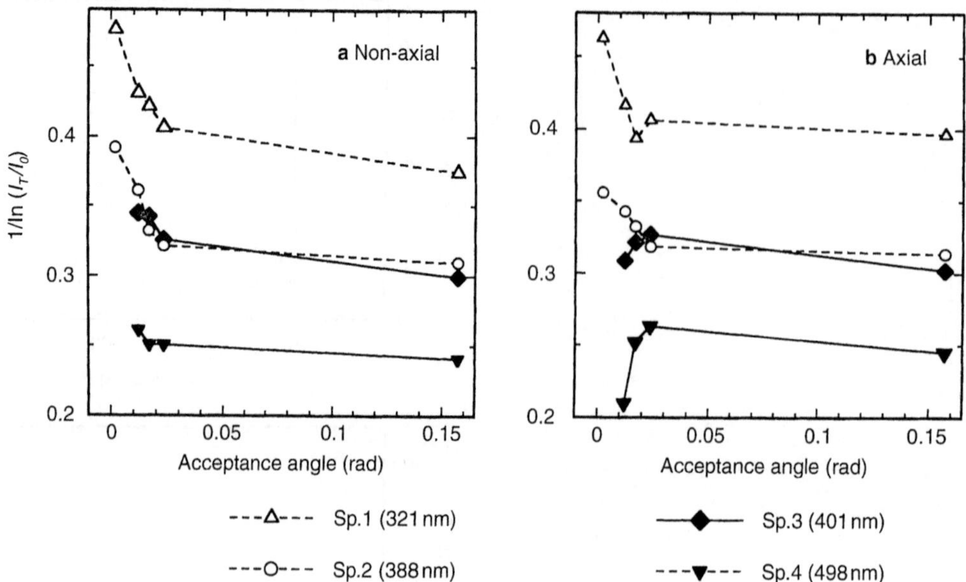

Fig. 3.23. Relation between $1/\ln(I_T/I_0)$ and t for α-Fe_2O_3 as a function of the acceptance angle β

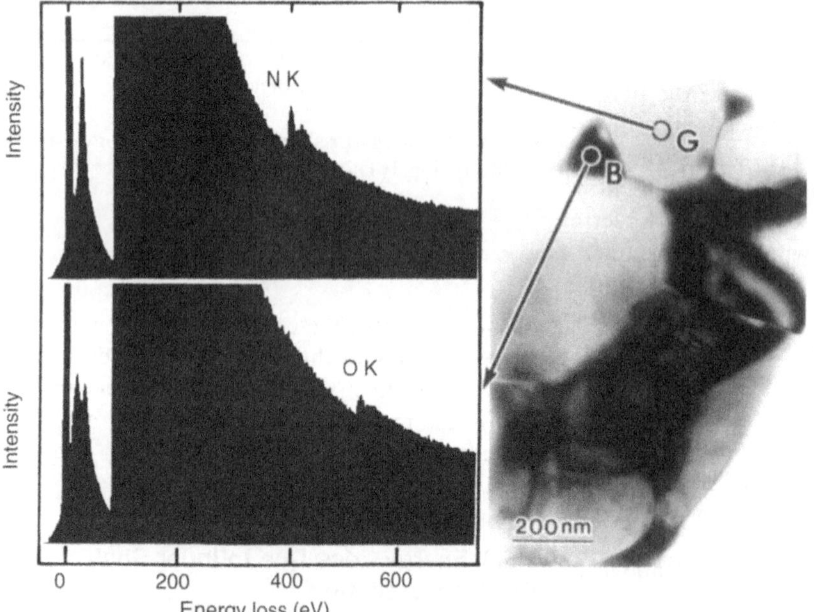

Fig. 3.24. Example of qualitative analysis in SIAlON

from an EDS spectrum (see Fig. 4.7), the nitrogen K-edge and oxygen K-edge appear on a high background, so it is necessary to remove the background for quantitative elemental analysis. The net signal intensity of an element can be obtained by subtracting background I_B, which is approximated with

$$I_B = C \cdot E^{-r} \tag{3.31}$$

where E is loss energy; and C and r are constants. On the other hand, the intensity of the core loss I_A for element A involved in the area is theoretically given with Eq. 3.6 and the intensity of the incident beam I_T, as

$$I_A = n_A t \sigma_A(\beta, \Delta E, E_0) I_T \tag{3.32}$$

where n_A and t are the number of atoms in a unit volume for element A and specimen thickness, respectively; and $\sigma_A(\beta,\Delta E,E_0)$ is the partial cross section of element A for acceptance angle β, energy window width ΔE, and incident electron energy E_0. The scattering cross section is larger for the lower atomic number. Therefore, EELS provides higher sensitivity for lighter elements. For quantitative analysis with EELS, using the signal intensity detected and the scattering cross section calculated, the number of atoms in the area can be evaluated as

$$n_A = \frac{I_A}{I_T} \cdot \frac{1}{t \cdot \sigma_A(\beta, \Delta E, E_0)} \tag{3.33}$$

For one spectrum, by removing specimen thickness and incident electron intensity, the content ratio of two elements (A and B) is obtained as [21]

$$\frac{n_A}{n_B} = \frac{I_A}{I_B} \cdot \frac{\sigma_B(\beta, \Delta E, E_0)}{\sigma_A(\beta, \Delta E, E_0)} \tag{3.34}$$

Note that especially thin specimens are necessary for EELS studies, where plural scattering is negligible, as plural scattering directly disturbs the accuracy.

3.5.3.2 Fine Structure in Energy-Loss Spectrum Due to Inner Shell Excitation

As mentioned in Section 3.4, energy loss due to inner-shell excitation contains information about the electronic structure, such as the unoccupied density of states. Therefore, much effort has been devoted to analysis of fine structures in various materials. In this section the K-edges of boron, carbon, and oxygen and an L-edge of transition metals are presented as typical examples of fine structures in energy-loss spectra due to inner-shell excitation. Energy loss due to the 1s → 2p transition is observed in the former, and energy loss due to the 2p → 3d transition is observed in the latter. In the region from the edge to 50 eV higher than the edge, the unoccupied density of states is strongly reflected in the spectrum; this is called ELNES (energy-loss near-edge structure). On the other hand, in energy-loss regions where the energy is 50 eV higher than that at the edge, the environmental state of the atom excited (i.e., information on atomic distance) is strongly reflected. It is called EXELFS (extended energy-loss fine structure) to distinguish it from ELNES (see Fig. 3.30). EXELFS corresponds to EXAFS (extended X-ray absorption fine structure) in X-ray absorption spectra.

K-Edges of Boron and Carbon. Figure 3.25 shows dependence of the boron K-edge on the scattering angle [22] obtained from boron nitride (BN), which has a hexagonal structure (Fig. 3.26). Sharp peaks called π^* and σ^* are observed in the boron K-edge at an energy of 188 eV and 194 eV, respectively. These peaks correspond to the excitation from the inner-shell (1s) to the unoccupied conduction bands π^* and σ^*, respectively.[2]

The π^* band corresponding to the $2p_z$ orbitals of boron and nitrogen is localized along the c-axis; and the σ^* band corresponding to the $2p_{x,y}$ orbitals of boron and nitrogen is extended on the c-plane. Therefore, as shown in Fig. 3.27, the excitation process with a component of the scattering vectors parallel to the c-axis (Fig. 3.27a), (i.e., excitation to π^*) is detected selectively when acquiring a spectrum around the transmitted beam. On the other hand, a component of the scattering vectors parallel to the c-plane (Fig. 3.27b), (i.e., excitation to σ^*) is detected by acquiring a spectrum at a position far from the transmitted beam. In Fig. 3.25 the spectra are obtained at the position of the transmitted beam and at the positions of $\frac{1}{4}d$ and $\frac{1}{2}d$, which correspond to one-fourth and one-half the distance between the transmitted beam and the 100 reflection ($1/d$, where d is 0.217 nm). The change from the π^* transition to the σ^* transition for a change in the scattering angle is clearly shown, as an acceptance angle width (angle resolution) was set to be smaller than $\frac{1}{50}$d.

Concerning π-bonding and σ-bonding of carbon, it is well known that carbon has various crystal structures, such as in diamond, graphite, and amorphous states. Diamond has a diamond-type crystal structure (the same as silicon), as shown in Fig. 3.28a; and a carbon atom bonds with the four atoms around it (four-coordination). C–C bonding, called σ-bonding, is strong and stable. The σ^* at 291 eV corresponding to σ-bonding is observed on a spectrum of diamond (Fig. 3.29).

Graphite has a graphite-type structure consisting of six-member ring layers, as shown in Fig. 3.28b. Here C binds with three atoms (three-coordination). This C–C bonding is also σ-bonding, and the bonding energy is the same as for diamond. It must be noted that carbon basi-

[2] The bonding orbitals occupied by electrons are called π and σ; and the antibonding orbitals, which are at higher energy and not occupied by electrons, are called π^* and σ^*.

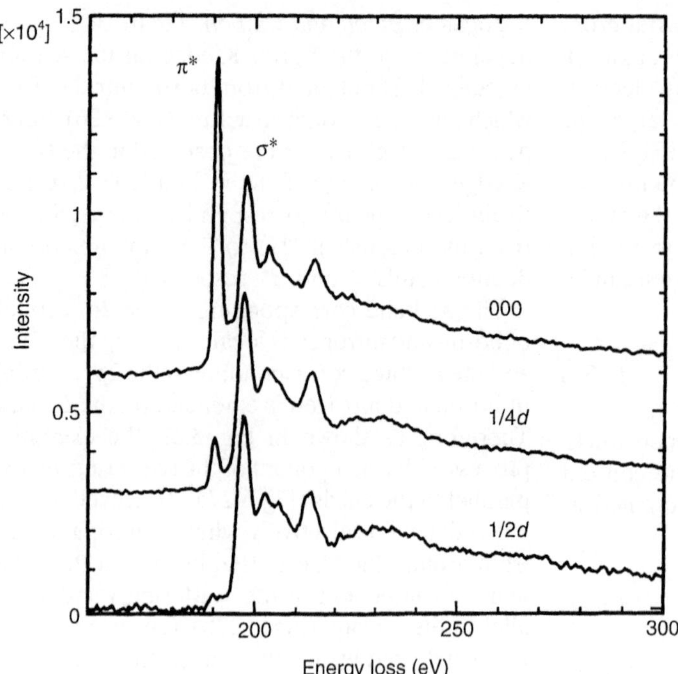

Fig. 3.25. Dependence of boron K-edge on the scattering angle in boron nitride

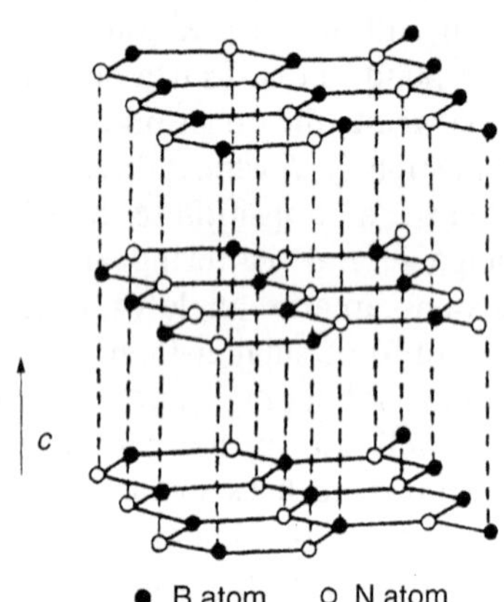

Fig. 3.26. Hexagonal structure of boron nitride

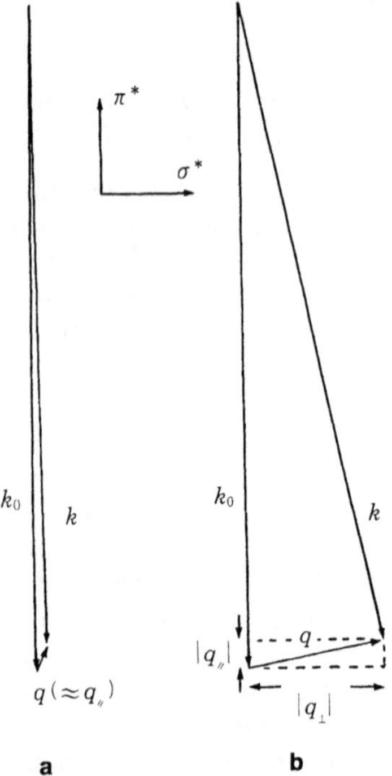

Fig. 3.27. Scattering vector for hexagonal boron nitride. **a** q parallel to the c-axis. **b** q perpendicular to the c-axis

cally has four bonding sites, and so one site remains open during three-coordination. Thus one is localized outside the six-member ring and binds with an atom in the neighboring layer. This is called π-bonding. Thus, in the carbon K-edge, the peak at 284 eV corresponding to π-bonding is observed in the spectrum of graphite, similar to the hexagonal boron nitride (*h*-BN) noted above. Moreover, a σ^* peak is observed at 291 eV on the K-edge (Fig. 3.29). For amorphous carbon, a small peak is visible at the π^* position of the spectrum. The peak shows that a small amount of micro-

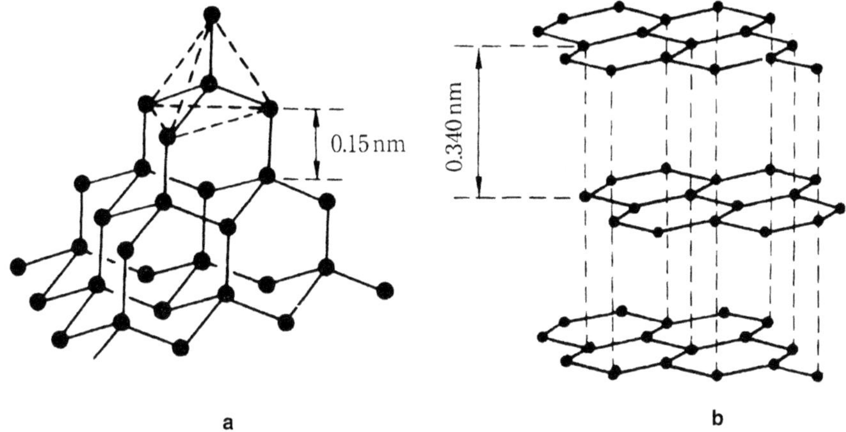

Fig. 3.28. Structures of diamond **a** and graphite **b**

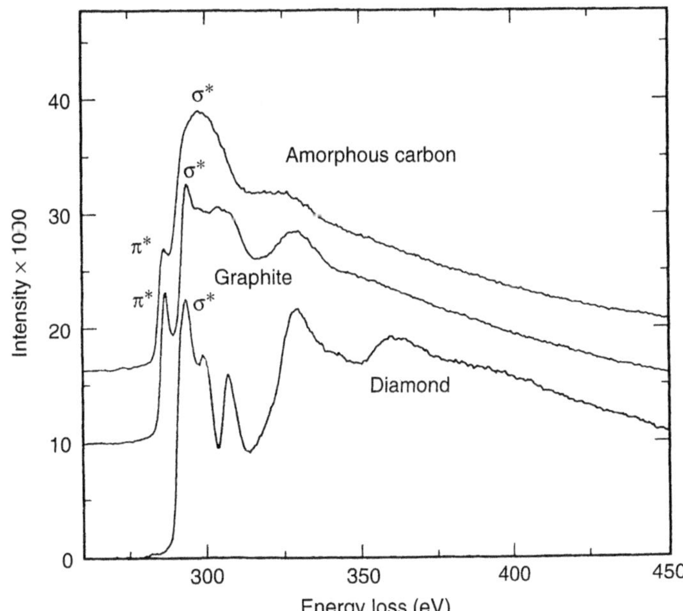

Fig. 3.29. Energy-loss spectra observed from amorphous carbon, graphite, and diamond

crystals having three-coordination is included. The peak at position σ^* is broad, indicating that the interatomic distance of σ-bonding is not constant. It has been reported that the π^* peak is observed from C atoms with three-coordination located at grain boundaries or lattice defects in diamond [23].

Figure 3.30 shows analysis of higher energy-loss region (EXELFS) of a carbon K-edge of graphite [24]. Figure 3.30a shows the spectrum obtained by removing the background from the original spectrum, and Fig. 3.30b is the spectrum obtained by the deconvolution operation with the spectrum of a low energy-loss peak region including the zero-loss peak and the plasmon-loss peak for removing the influence of the plural scattering. Figure 3.30c shows the spectrum of an oscillation component obtained by subtracting the nonoscillation component from that in Fig. 3.30b. A radial distribution function (RDF) (Fig. 3.30d) can be obtained by making the Fourier transform on the spectrum in Fig. 3.30c. A main peak at 0.14 nm in Fig. 3.30d shows good agreement with the interatomic distance (0.14 nm) of carbon atoms in graphite.

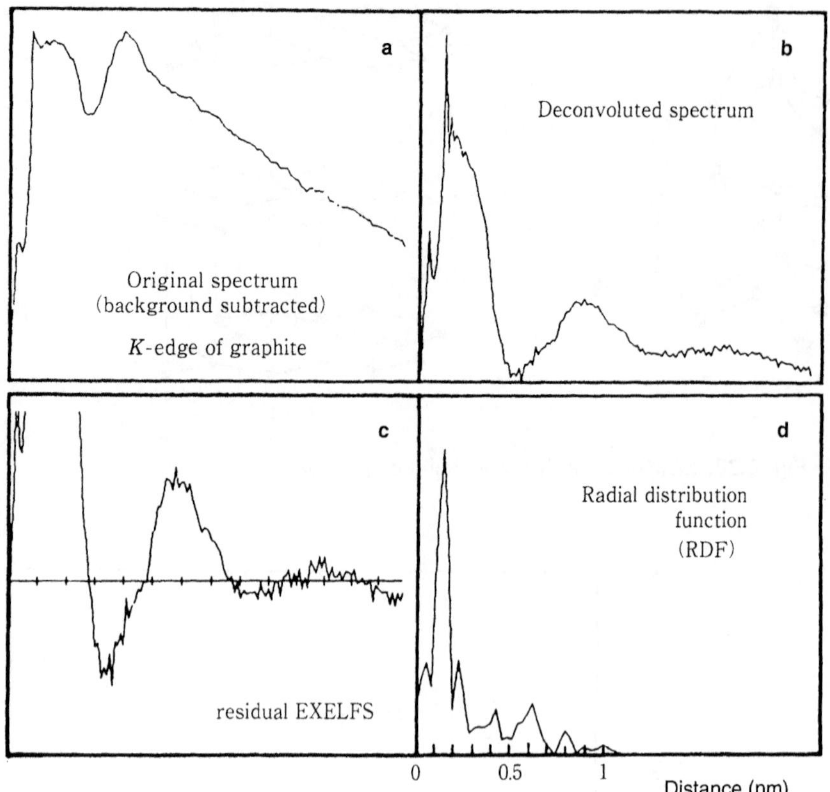

Fig. 3.30. Analysis on extended energy-loss fine structure (EXELFS) of the carbon K-edge of graphite

Because data for the analysis of EXELFS require high statistical accuracy, the analysis is usually limited to K-edges of light elements.

Oxygen K-Edge. Concerning Cu-based oxides (superconducting oxide), Fig. 3.31 shows the K-edge spectra (1s → 2p) of oxygen in superconducting oxide $La_{2-x}Sr_xCuO_4$ [25]. By substituting Sr^{2+} for La^{3+} the content of holes (carriers of the superconductor) in the material is increased. With the increase in hole content, the height of the shoulder at around 528 eV increases. The result indicates that the holes are mainly located at oxygen positions. Hence, the electronic structure of these materials is interpreted with the density of states shown in Fig. 3.32.

Figures 3.33 and 3.34 show changes of an oxygen K-edge in manganese oxide $Bi_{1-x}Ca_xMnO_3$ for the composition change and the phase transformation, respectively [26]. The manganese oxides have attracted much attention for colossal magneto resistance. It is reported that the holes introduced by substituting Ca^{2+} for Bi^{3+} are located at Mn-3d orbitals. The relation of the height of the peak (Fig. 3.35, A) to the hole content shown in Fig. 3.35 is thought to result

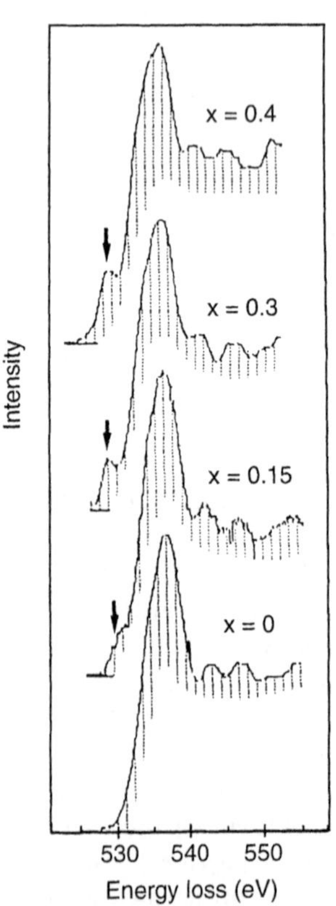

Fig. 3.31. Oxygen K-edge in $La_{2-x}Sr_xCuO_4$

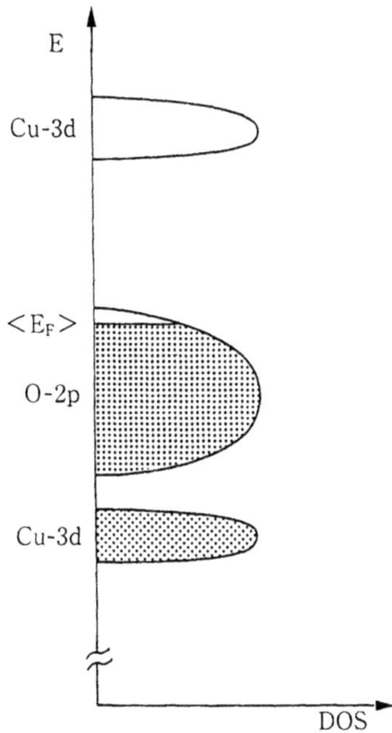

Fig. 3.32. Density of states in Cu-O

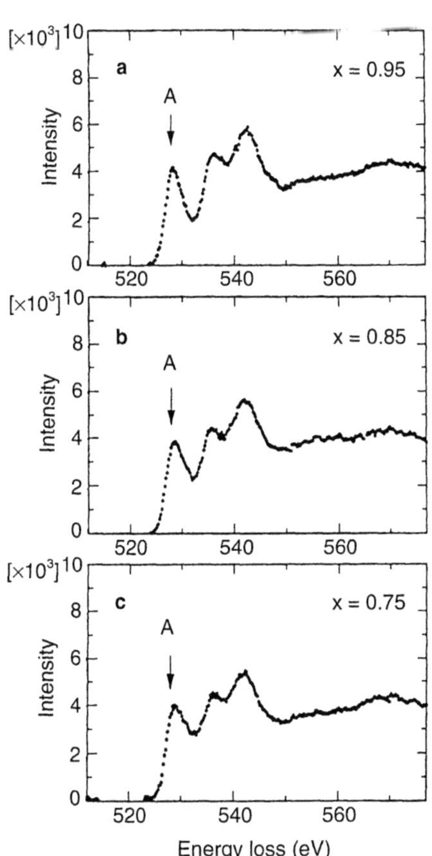

Fig. 3.33. Oxygen K-edge of $Bi_{1-x}Ca_xMnO_3$

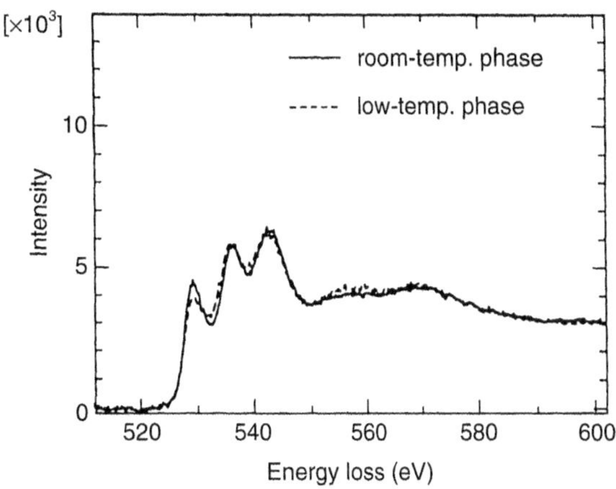

Fig. 3.34. Change of oxygen K-edge of $Bi_{0.2}Ca_{0.8}MnO_3$ for phase transformation

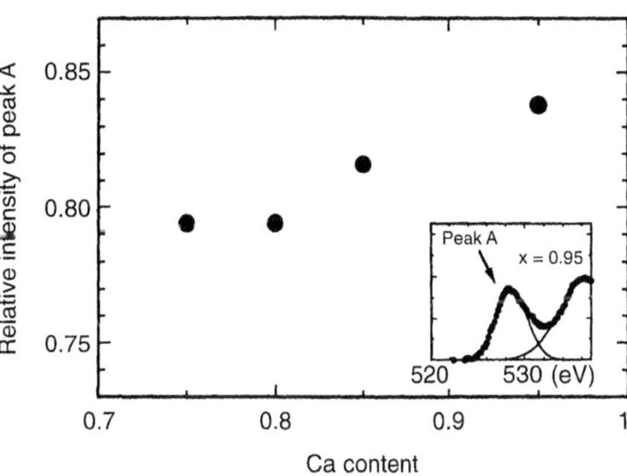

Fig. 3.35. Change of oxygen K-edge (peak A in Fig. 3.33) of $Bi_{1-x}Ca_xMnO_3$ for the composition change

from hybridization of the Mn-3d and oxygen-2p orbitals. On the other hand, the height of the peak is lower owing to the phase transformation with a decrease in temperature. It is thought that the effect of the so-called charge ordering in Mn^{3+} and Mn^{4+} produces the lattice strain resulting in decreased hybridization of Mn-3d and oxygen-2p orbitals.

L-Edge of Transition Metals. In transition metals with partly filled d-bands, sharp peaks or edges (so-called white lines) appear in an energy-loss spectrum owing to the transition (p → d), which satisfies the selection rule. Occupation of electrons in the d-bands has been estimated from measuring the intensity of white lines [27]. To detect a change in electronic structure from the energy-loss spectrum, it is important to compare the spectra in thin specimens of the same thickness. The detection of L-edge change in $Ti_{50}Ni_{48}Fe_2$ due to the phase transformation is presented. Figure 3.36a,c shows L-edges of Ti and Ni, respectively. The change in heights of the L-edges due to the martensitic phase transformation (parent → martensite) is clearly observed in the enlarged spectra in Fig. 3.36,b,d. Here, L_2 and L_3 indicate white lines corresponding to the transitions from 2p1/2 and 2p3/2 to 3d (for notations such as L_2 2p1/2, see Section 4.1). The martensitic phase transformation can be confirmed from the microstructure change shown in Fig. 3.37. Energy-loss spectra were measured in the area indicate by "X" in Fig. 3.37. It is seen that the heights of the Ti $L_{2,3}$-edges decrease owing to the phase transformation, whereas those of the Ni $L_{2,3}$-edges increase. To confirm the change, the spectra were observed in the cycle parent → martensite → parent, as shown in Fig. 3.38. It is concluded that the unoccupied density of states in Ti decreases owing to the martensitic phase transformation, whereas that of Ni increases [28]. For quantification of the white line intensity, the integrated intensity of the peak (I_A) should be normalized with the background intensity (I_B), which is situated far from the edge (Fig. 3.39). Table 3.2 shows the difference in the normalized intensities due to the phase transformation $[(I_A/I_B)_M - (I_A/I_B)_P]$ observed in different areas in two specimens (#1 and #2). It is seen that the tendencies of the electronic structure change are confirmed in different areas.

Figure 3.40a shows Ti L-edges observed at various regions in $Ti_{50}Ni_{48}Fe_2$. The spectra are normalized by the background heights below the L-edges. It is seen that the height of the L-edge is large for thin crystal regions. In Fig. 3.40b, the peak in the L-edge/background (P/B) ratio is plotted as a function of specimen thickness. The

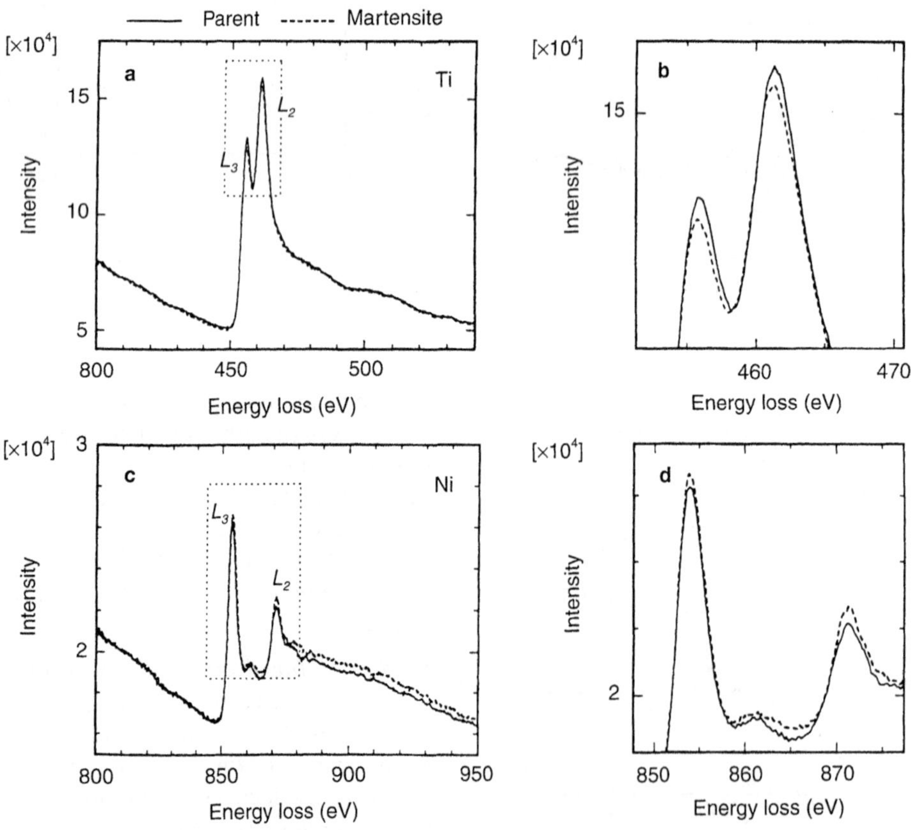

Fig. 3.36. Change of L-edges of Ti **a** and Ni **c** in $Ti_{50}Ni_{48}Fe_2$ for phase transformation. **b**, **d** Enlarged spectra of **a** and **c**

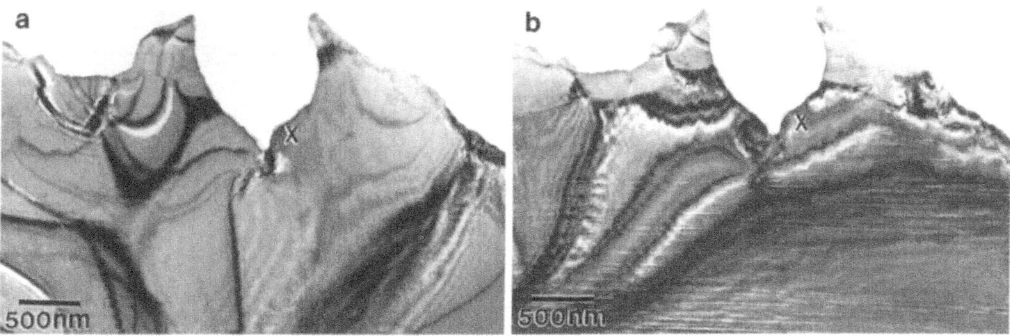

Fig. 3.37. Microstructure change in $Ti_{50}Ni_{48}Fe_2$ due to martensitic phase transformation. **a** Parent. **b** Martensite

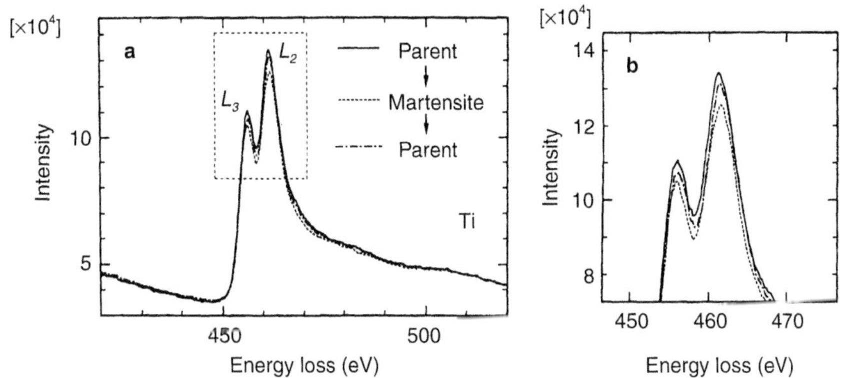

Fig. 3.38. **a** Change of Ti-$L_{2,3}$ edges due to phase transformation in the cycle parent → martensite → parent. **b** Enlarged spectra in **a**

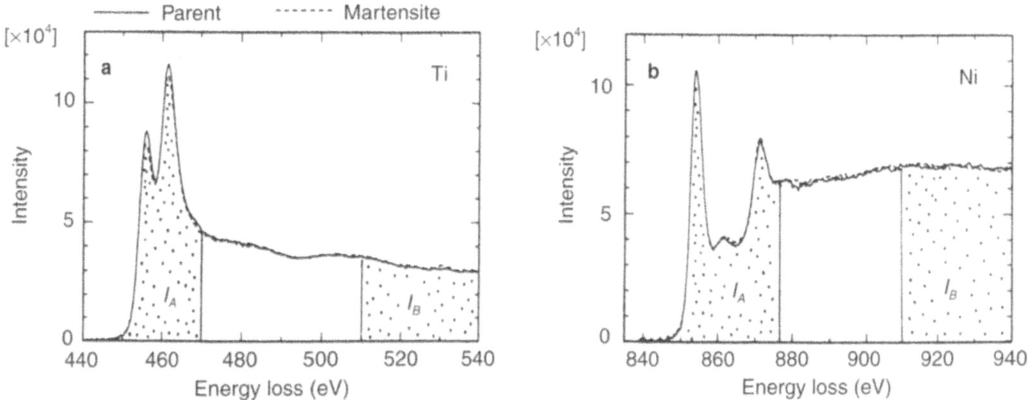

Fig. 3.39. Integrated intensity of the peak (I_A) and background intensity (I_B) in the energy-loss spectra of Ti **a** and Ni **b**

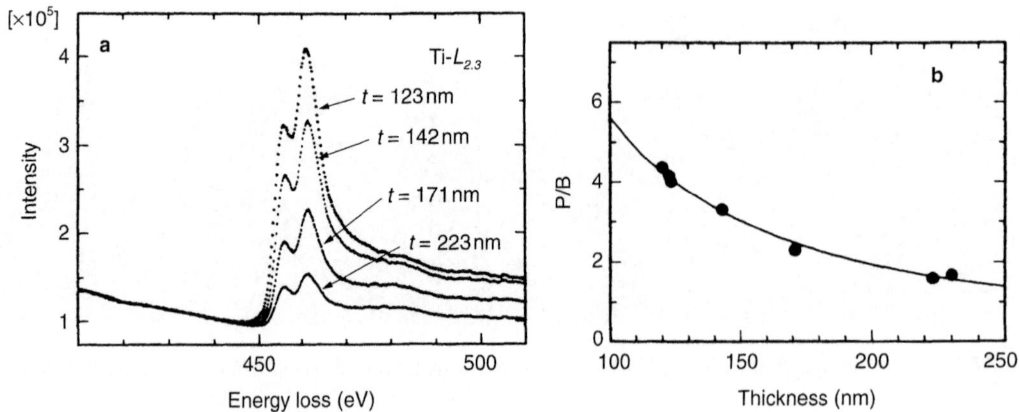

Fig. 3.40. **a** Change of the Ti L-edge observed in $Ti_{50}Ni_{48}Fe_2$ of various thicknesses. **b** Peak/background (P/B) ratio of the Ti L-edge as a function of specimen thickness

Table 3.2. Intensity ratio (I_A/I_B) of parent (P) and martensite (M) phases and their difference.

		$(I_A/I_B)_P$ Parent	$(I_A/I_B)_M$ Martensite	$(I_A/I_B)_M$-$(I_A/I_B)_P$ Martensite-Parent
# 1-1	Ti	1.33	1.27	-6×10^{-2}
	Ni	0.682	0.687	$+5 \times 10^{-3}$
# 1-2	Ti	1.37	1.31	-6×10^{-2}
	Ni	0.832	0.832	0
# 1-3	Ti	1.57	1.51	-6×10^{-2}
	Ni	0.721	0.728	$+7 \times 10^{-3}$
# 2-1	Ti	0.899	0.836	-6×10^{-2}
	Ni	0.645	0.653	$+8 \times 10^{-3}$
# 2-2	Ti	0.959	0.946	-1×10^{-2}
	Ni	0.706	0.707	$+1 \times 10^{-3}$

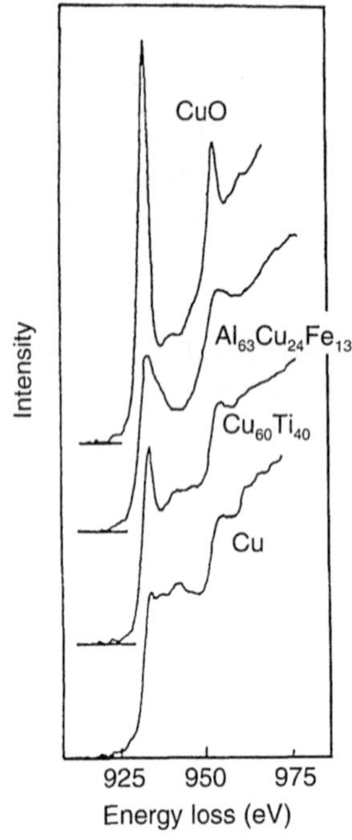

Fig. 3.41. Cu-$L_{2,3}$ edges in Cu and Cu-based compounds

P/B ratio decreases gradually with the increase in specimen thickness. Although a high P/B is obtained at a thickness of less than 150nm, it should be noted that the thin film effect is enhanced at thicknesses less than 100nm; and the behavior of the martensitic transformation in such a thin region deviates from that of a thick area or a bulk specimen.

Figure 3.41 shows the Cu L-edges in Cu and Cu-based compounds [29]. For Cu, no sharp peaks at the L-edge are observed because Cu-3d bands are filled with 10 electrons. On the other hand, in compounds such as CuO, two sharp peaks or white lines are observed at the L_3 and L_2 positions in the spectrum. It is thought that for formation of the compound some electrons in the Cu-3d bands are supplied to oxygen and thus some of the Cu-3d bands are vacant. For $Al_{63}Cu_{24}Fe_{13}$ and $Cu_{60}Ti_{40}$, relatively small peaks are observed at the $L_{2,3}$-edges. In the Cu-based compounds containing other transition metals, it is thought that there is some charge transfer between the 3d bands of Cu

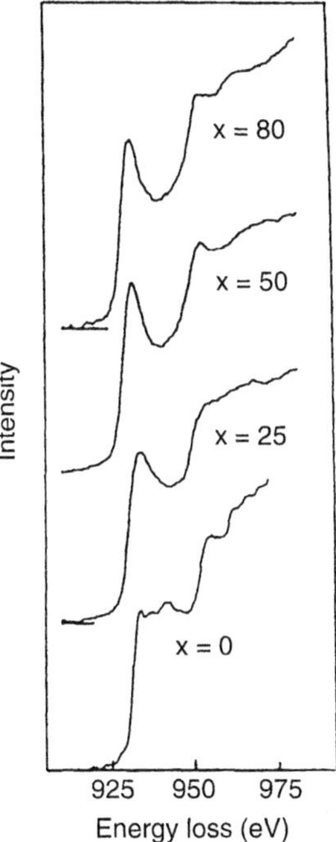

Fig. 3.42. Cu $L_{2,3}$-edges of $Cu_{100-x}Al_x$

and other transition metals, so some of the Cu-3d bands are vacant. Compared with the spectrum of $Cu_{60}Ti_{40}$, the widths of the $L_{2,3}$-edges in $Al_{63}Cu_{24}Fe_{13}$ are wide. The small wide peaks are thought to result from hybridization of Cu-3d bands and Al-s,p bands, the latter of which are much wider than 3d bands. Broad peaks in Cu $L_{2,3}$-edges are found to increase monotonically with the increase in Al content in $Cu_{100-x}Al_x$ shown in Fig. 3.42.

3.6 Principles and Application of Energy Filtering

3.6.1 Electron Optics of Energy Filter

Figure 3.43a shows an optical ray diagram of an omega-type filter. There is a crossover point at the entrance point of the spectrometer and one at its exit point. An image formed by the intermediate lens system at the prelenses of the filter is an entrance image, and the image formed in front of the exit crossover point is the exit image. In the imaging mode, the focus of the intermediate lens is adjusted so it is on the image formed by the objective lens; and thus both entrance and exit images are electron microscopic images. In the diffraction mode, the focus of the intermediate lens is adjusted so it is on the diffraction pattern (back focal plane) formed by the objective lens; so both entrance and exit images are electron diffraction patterns.

Figure 3.43b shows the energy dispersion of the omega-type filter. The incident electrons passing through the first spectrometer with energy dispersion enter the second spectrometer. At the back of the second spectrometer, the energy dispersion is canceled. The third spectrometer produces the energy dispersion again; then the non-energy-dispersion plane, the so-called *achromatic plane*, appears in the fourth spectrometer. Finally, large energy dispersion is produced under the fourth spectrometer.

The spectrometer is designed to make the exit plane equal to the final energy dispersion plane, so the sharply focused energy-loss spectrum is formed on this energy dispersion plane. On the other hand, it is also designed to make the exit image at the achromatic plane of the fourth spectrometer, and the image is thus called the achromatic image.

When the focus of the projector lens located below the spectrometer is set to be on the energy dispersion plane, an energy-loss spectrum can be seen on the screen. When the focus of the projector lens is set to be on the achromatic image plane, an electron microscopic image is observed. In this observation mode, an energy-filtered image is obtained by selecting the electrons with a specific energy loss, with the energy-selecting slit located at the dispersion plane. When the slit is retracted from the dispersion plane, an unfiltered image is obtained.

In this way, the focus of the projector lens is set to be at the achromatic image plane, so an achromatic image without energy spread is obtained even if the wide slit is inserted and if the slit is retracted. The principle of these optics is the same as that for other types of energy filter.

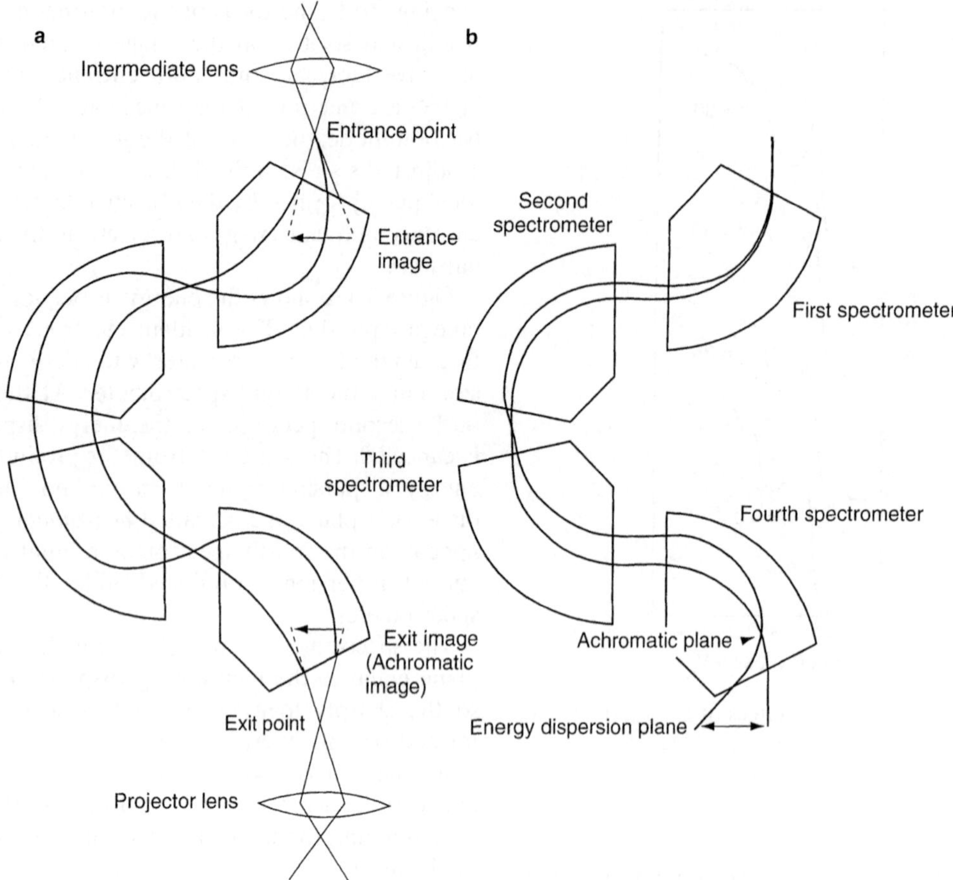

Fig. 3.43. Optical ray diagram **a** and energy dispersion **b** of an omega-type filter

3.6.2 Energy Filters and Their Features

The energy filters installed on electron microscopes have recently been categorized as being of two types (Fig. 3.46). One is the so-called in-column type, in which the energy filter is inserted in the microscope column. The omega type [30] and the Castaing-Henry type [31] are examples of the in-column type. The other is the so-called post-column type, where the filter is installed at the bottom of an electron microscope (i.e., under the camera chamber). Typicalally for the post-column type, a sector-type energy filter is popular [32,33]. In both cases, energy-loss spectra can be observed on the screen or the monitor, and electrons with a specific energy can be selected by the energy-selecting slit to record energy-filtered images and diffraction patterns. In a microscope with the in-column-type filter, all the recording systems (e.g., photo-film, imaging plate, television camera, slow-scan CCD camera) can be used. On the other hand, only the slow-scan CCD camera can be used in microscopes with the post-column-type filter. It is rather difficult to obtain diffraction patterns for a wide range of scattering angles with the post-column-type filter. However, because it is rather easy to install this filter in various microscopes, it is expected to apply especially to high-resolution filtered-image observations. The features of each filter are described below.

3.6.2.1 Castaing-Henry Type Energy Filter

The Castaing-Henry type energy filter was developed by Castaing et al. in 1962 [31] and commercialized by the Zeiss Company in 1986. Its production was stopped in 1994. The filter is inserted between two intermediate lenses. The trajectory of the electrons is bent by 90°, and their energy is dispersed with the magnetic spectrometer with a triangular shape. The electrons are then reflected by an electrostatic mirror located by the side of the magnetic spectrometer. They enter the magnetic spectrometer again, and energy

Aberration of an Energy Filter and Isochromaticity

When the energy filter to be used has no aberration, an energy-loss spectrum without distortion forms on the energy dispersion plane, as shown in Fig. 3.44. In practice, there are always some aberrations, so a rectangular spectrum for a spectrometer with single focusing has distortion (see Fig. 3.48, below). Figure 3.45a shows a energy-loss spectrum with double focusing affected by an aberration including a second-order aberration. When an equienergy axis is not perpendicular to the energy-loss axis, a distortion tends to appear. When the distorted spectrum is selected by a slit with a parallel line shape prepared mechanically, electrons with different energies are selected especially for a relatively large entrance angle (γ). In the case of no aberration, the energy at P is equal to the incident energy (i.e., no energy loss); with aberration, the energy of this point corresponds to the energy loss ($-\Delta E$) where the energy-loss axis and the distorted equienergy axis intersect. If an energy-filtered image is obtained under this condition, the filtered image, whose energy is different at the center and the outside, is obtained (Fig. 3.45b). The shaded region corresponds to the image consisting of the electrons with energy less than $-\Delta E$. In general, the larger the angle of the entrance aperture, the larger is the aberration of the spectrometer. The homogeneity of the energy in the limited area of the filtered image is called isochromaticity.

In practice, an objective aperture, a selected area aperture, and an entrance aperture of the spectrometer are inserted for reducing the entrance angle to suppress the aberration (see Fig. 3.47, below). Moreover, if the width of the energy-selecting slit is increased enough, the energy difference in the filtered image (non-isochromaticity) can be neglected.

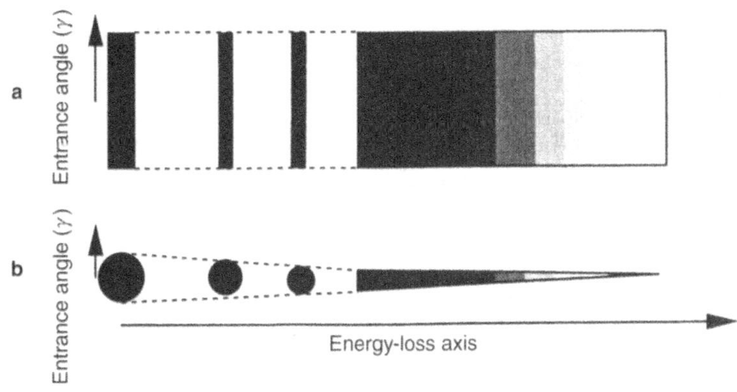

Fig. 3.44. Energy-loss spectra without distortion. **a** Single focusing. **b** Double focusing

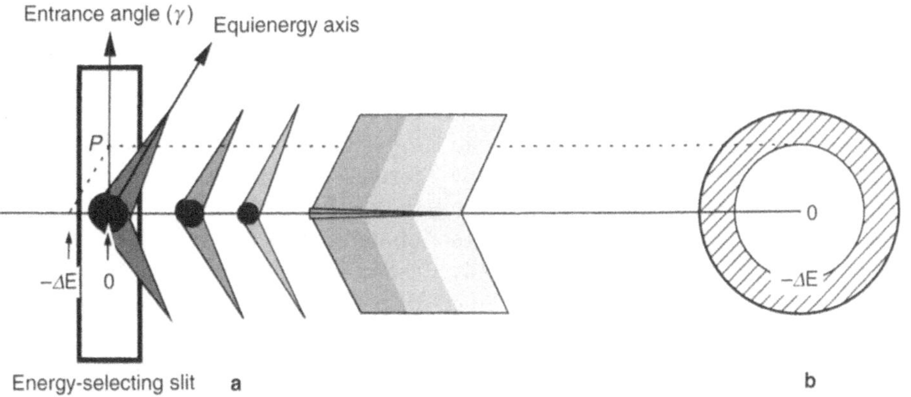

Fig. 3.45. Spectrum of a spectrometer with double focusing affected by aberration **a** and a resultant filtered image **b**

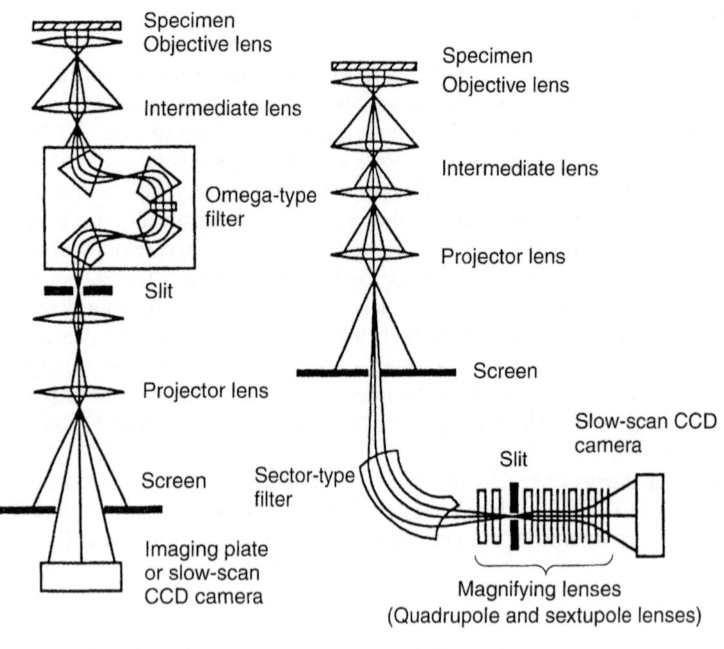

Fig. 3.46. Two types of energy filter

dispersion forms on the microscope axis. By selecting electrons with a specific energy using an energy-selecting slit and adjusting the focus of the intermediate lens below the filter at the achromatic image plane, an energy-filtered image is obtained on the screen. Because voltage the same as the accelerating voltage should be applied to the mirror electrode, there are some technical limitations during its application to high voltage electron microscopes.

3.6.2.2 Omega-Type Energy Filter

The omega-type energyy filter (Fig. 3.43) was developed by Zanchi et al. in 1975 [30] and commercialized by the Zeiss Company in 1991 and the JEOL Company in 1997. Basically, its principle is the same as the Castaing-Henry type, but instead of the electrostatic mirror additional magnetic spectrometers with the magnetic field opposite to that of the preexisting magnetic spectrometers are employed. Because it is not necessary to supply high voltage to the magnetic spectrometer, the spectrometer has the advantage of application to high voltage microscopes. The filter consists of four (or three) spectrometers, and the shape of the electron trajectory is just like the Greek letter Ω: hence it is called an omega (Ω)-type filter. Figure 3.47 shows an energy-loss spectrum and its intensity distribution obtained with this filter. Because the imaging plate can be used in addition to a slow-scan CCD camera in the microscope with this filter, it is possible to record the energy-loss spectrum with high precision utilizing more than 3000×3000 pixels. Because the height of the filter or the distance from the entrance point to the exit point is rather long, it has the disadvantage of making the microscope column long.

3.6.2.3 Sector-Type Energy Filter

The sector-type energy filter (Fig. 3.46b) was developed by Ajika et al. in 1983 [34] and commercialized by the Gatan Company in 1992. It was developed on the basis of a conventional spectrometer formerly installed at the bottom of a camera chamber in an electron microscope. The energy-selecting slit is inserted at the energy dispersion plane, and electrons with a specific energy can be selected. The final filtered image is formed by utilizing the magnifying lenses behind the slit. Figure 3.48 shows an energy-loss spectrum obtained with this filter. The spectrum has a line shape because the spectrometer has line-focusing dispersion (so-called single focusing). The white band corresponding to so-called zero loss is slightly bent. This is due to the second-order aberration of the spectrometer. Figure 3.49 shows energy dispersion in an electron diffraction pattern. In the systematic reflections along the vertical direction, each reflection has energy dispersion parallel to the horizontal line. Because the sector-type spectrometer generates a large distortion of the image, the distortion is usually corrected by

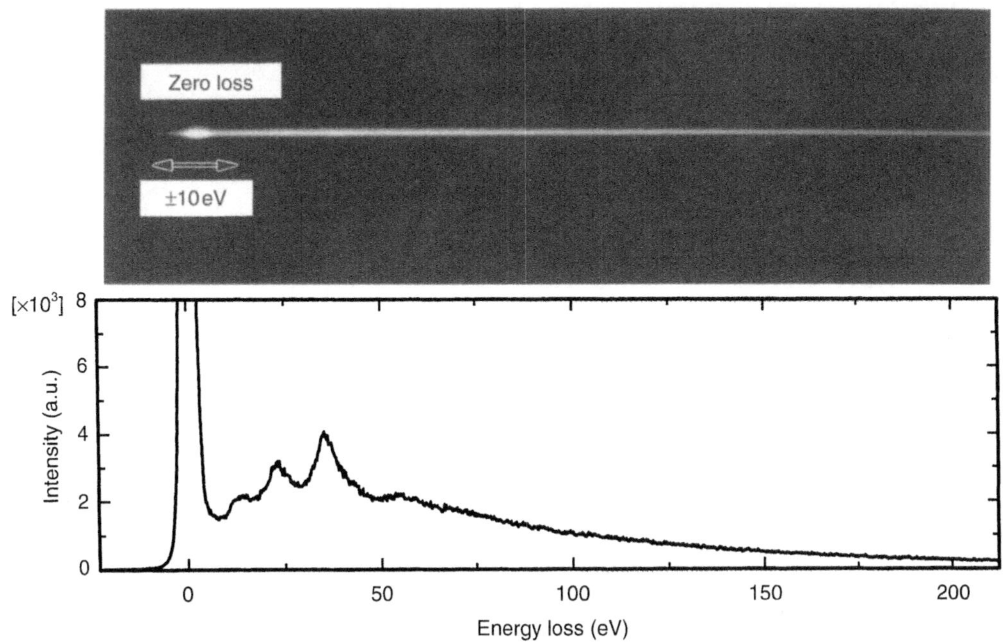

Fig. 3.47. Energy-loss spectrum and its intensity distribution obtained with the omega-type energy filter

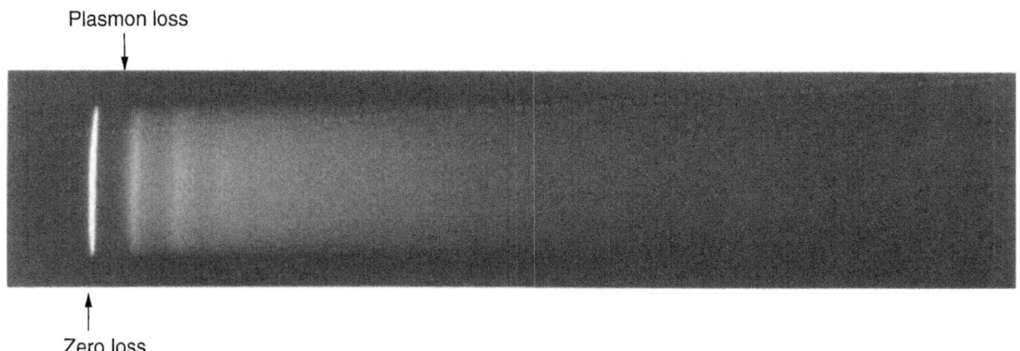

Fig. 3.48. Energy-loss spectrum obtained with the sector-type energy filter

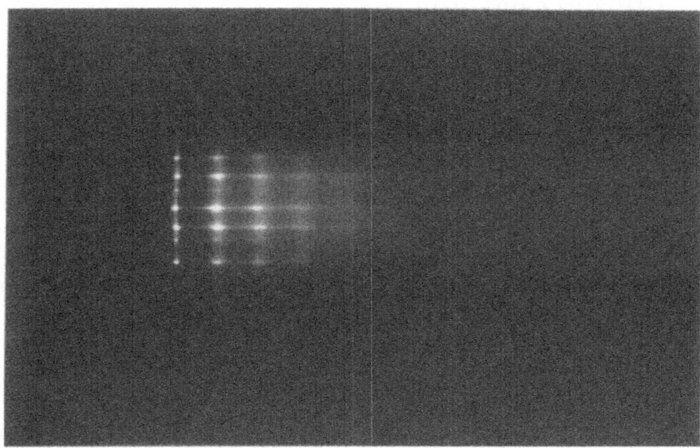

Fig. 3.49. Energy dispersion in an electron diffraction pattern observed with the sector-type energy filter

quadrupole and sextupole lenses. Only the central part of a magnified image is selected for observation, so limitation of the field of view is a disadvantage in this filter. On the other hand, the filter has the advantage that it can be installed on any microscope.

3.6.2.4 Gamma-Type Energy Filter

The gamma (γ)-type energy filter was developed by Taya et al. [35] and commercialized by the Hitachi Company in 1996. Its production was stopped in 1999. Its principle is basically the same as that for the omega-type, but the shape of the electron trajectory resembles the Greek letter γ. The filter is mechanically simplified by employing two magnetic spectrometers. Because it has the advantage of a small filter height, it is possible to obtain large energy dispersion even in a short column.

3.6.3 Analytical Technique and Application of Energy Filter

3.6.3.1 Energy Filtering and Spectrum Imaging

Two-dimensional analysis of an electron microscope image of a specimen with EELS is called "*energy filtering*." The microscope image is obtained by selecting electrons with a specific energy. Because the energy-loss spectrum generally has high background, processing the filtered image by subtracting the background component under the peak or edge of the element in the spectrum is necessary to obtain a net image showing elemental distribution (elemental mapping).

On the other hand, acquisition of the spectra at all pixels over the entire image is called "*spectrum imaging*." It takes a relatively long time to acquire spectra from a wide area. However, this method is quite useful for analyzing or processing the data in detail by various ways after acquisition, as whole spectra are acquired at each pixel of the area.

Elemental Mapping with EELS. The energy-filtered image of an element is obtained by selecting electrons in such a way as to locate the energy-selecting slit at the specific energy-loss value of the element. Because the elemental signal is superimposed on a high background in the spectrum, strictly speaking the energy-filtered image is an overlap of the elemental signal and the background. Therefore, the background should be subtracted from the filtered image to obtain a net elemental image. Background subtraction during energy filtering has been carried out for a long time [36]. Currently, two methods (described below) are widely used utilizing personal computers (PCs) because of their high performance.

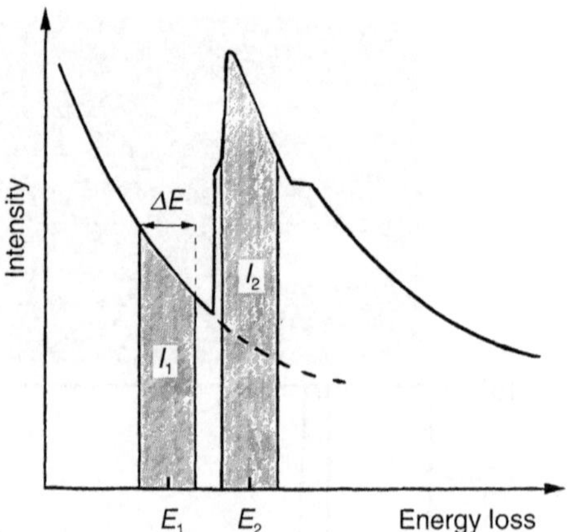

Fig. 3.50. Principle of the two-window method

For the *two-window (jump ratio) method* (see Fig. 3.50), two filtered images should be recorded in the same field of view. One is recorded at energy E_1 just below the energy-loss value corresponding to the specific edge (pre-edge). The other is recorded at E_2 just above the edge (post-edge). With these two images, the following operation is carried out for each pixel of the two images $I_1(x, y)$, $I_2(x, y)$ (Fig. 3.51).

$$I_S(x,y) = \frac{I_2(x,y)}{I_1(x,y)} \tag{3.35}$$

The resultant intensity $I_S(x, y)$ shows the elemental signal/background ratio (jump ratio); therefore, it is also called the jump ratio method. Although the intensitiy $I_S(x, y)$ does not indicate the content of the element quantitatively, the method is useful for observing the elemental distribution especially for thick specimens with accompanying high background and for specimens without uniform thickness.

For the *three-window method*, as shown in Fig. 3.52, three filtered images including two pre-edge images (I_1 and I_2) and a post-edge image (I_3) should be recorded. The background intensity can be fitted with the following equation (Eq. 3.31) as described above

$$I_B = C \cdot E^{-r}$$

and the background intensity is extrapolated by determining parameters C and r for each pixel.

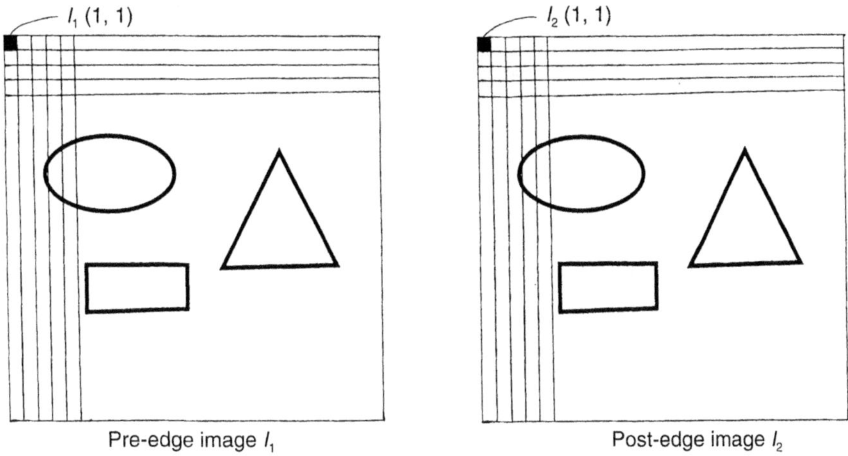

Fig. 3.51. Illustration showing pixels in I_1 and I_2

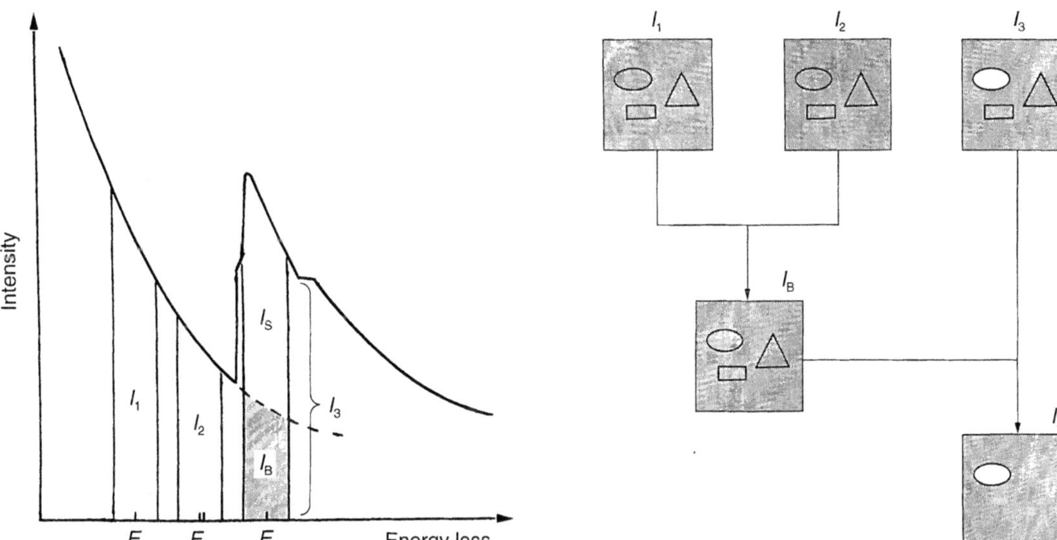

Fig. 3.52. Principle of three-window method

Fig. 3.53. Operation process for the three-window method

Thus, the background intensity $I_B(x, y)$ at the post-edge position can be obtained. Finally, the background is subtracted from the post-edge image $I_3(x, y)$ to obtain the net signal intensity $I_S(x, y)$ as

$$I_S(x,y) = I_3(x,y) - I_B(x,y) \qquad (3.36)$$

Figure 3.53 shows the process of the calculation. The resultant intensity $I_S(x, y)$ indicates the net elemental signal and thus shows the elemental content quantitatively. Before doing the processing, it is important to correct the image shifts among the three filtered images produced while obtaining the images.

Application of Elemental Mapping. Figure 3.54a shows elemental mapping on a nickel-based heat-resisting alloy (Inconel) with the two-window method [37]. The alloy is a precipitation hardening alloy consisting of precipitates of γ'-phase ($L1_2$-type structure) and γ''-phase ($D0_{22}$-type structure). Titanium (Ti) is included in the γ'-phase. The morphology of γ'-phase containing Ti is clearly observed with the two-window method on the Ti L-edge (Fig. 3.54b).

Figure 3.55 shows elemental mapping on ceramic material with the three-window method. The ceramic material contains SiC and SiN com-

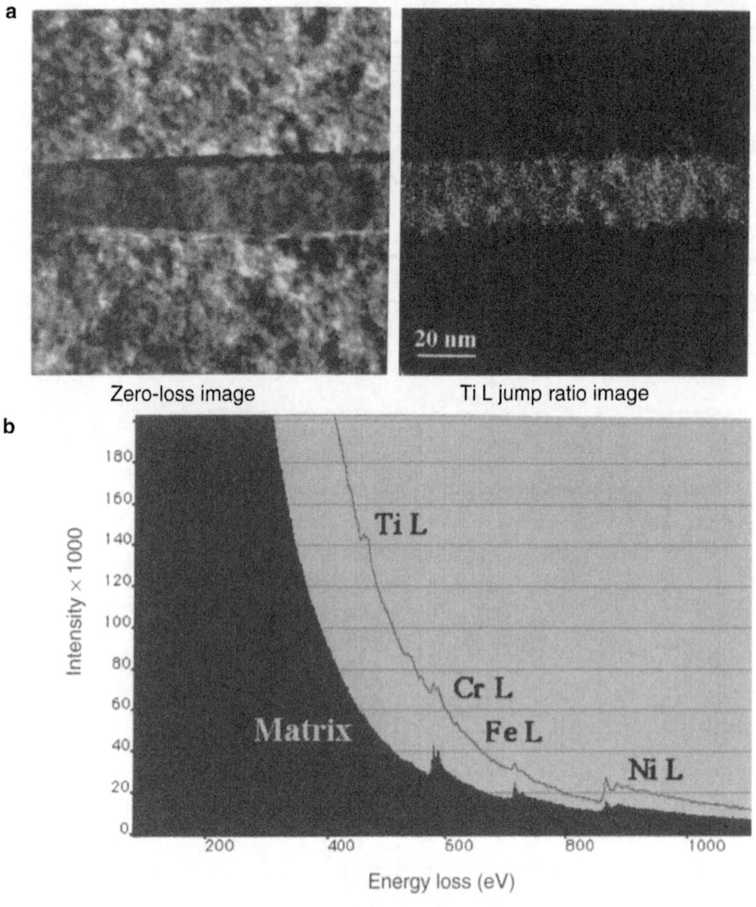

Fig. 3.54. **a** Elemental mapping on Inconel with the two-window method. **b** Its energy-loss spectrum

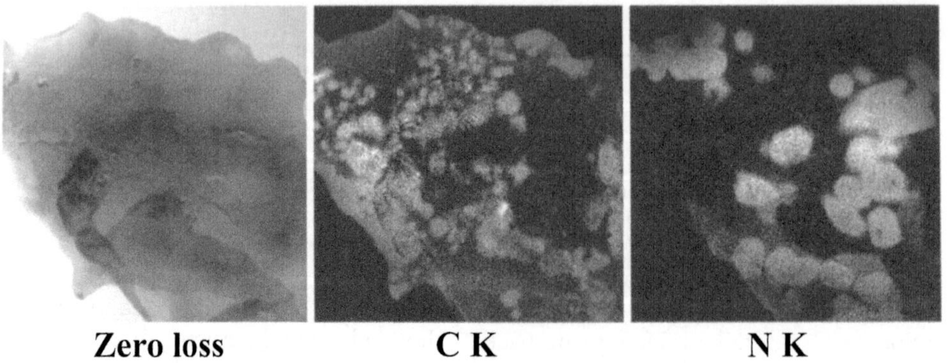

Fig. 3.55. Elemental mapping on SiN-SiC with the three-window method

pounds. By carrying out elemental mapping for carbon and nitrogen, the two Si compounds are clearly distinguished.

Spectrum Imaging. An electron micrograph provides information about the electron intensity in two dimensions with variables (X, Y). On the other hand, an energy-loss spectrum provides information on electron intensity in one dimension with a variable (ΔE). When energy-loss spectra of one-dimensional information are acquired for each point in a specimen area in two dimensions, the data eventually consist of the information in three dimensions with variables $(X, Y, \Delta E)$.

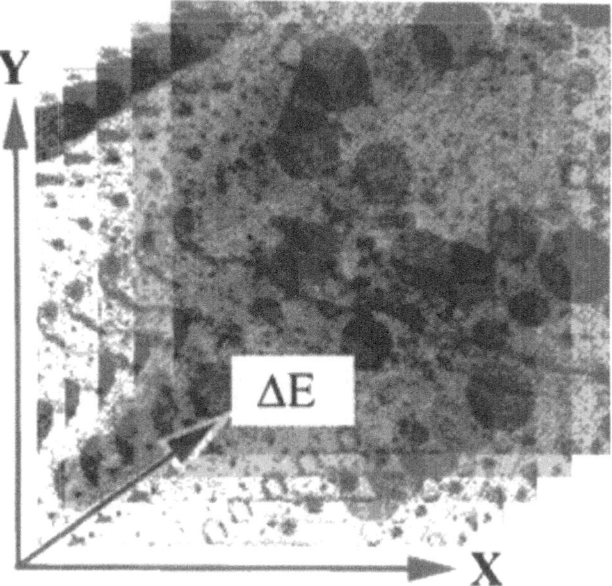

Fig. 3.56. Spectrum imaging in transmission electron microscopy (TEM) mode

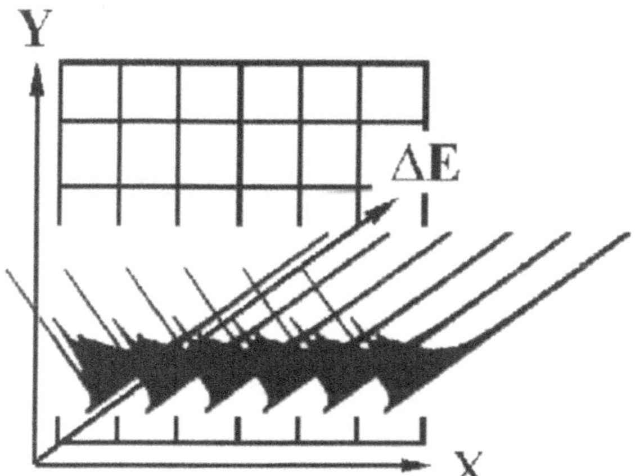

Fig. 3.57. Spectrum imaging in STEM mode

A set of three-dimentional data is convenient for analyzing the data in various ways. For example, the two-dimensional intensity distribution $I(X, Y)$ at $\Delta E = E_1$ gives an energy-filtered image, and the intensity distribution along axis ΔE at position (X_1, Y_1) gives an energy-loss spectrum. The technique for obtaining a set of three-dimensional data is called spectrum imaging. In the transmission election microscopy (TEM) mode, a series of energy-filtered images are acquired as a function of the energy-loss value, as shown in Fig. 3.56, for spectrum imaging. On the other hand, an energy-loss spectrum is acquired for each pixel of the specimen area in the scanning TEM (STEM) mode, as shown in Fig. 3.57.

3.6.3.2 Evaluation of Diffraction Patterns with Energy Filtering

Effect of Inelastically Scattered Electrons on Electron Diffraction Patterns. In addition to the dynamic diffraction effect, the background due to the inelastically scattered electrons is one the problems for quantitative analysis of electron diffraction patterns. The main components of the inelastic scattering are thermal diffuse scattering and plasmon scattering. When the independent vibration of constituent atoms in a specimen is assumed (Einstein model), the intensity of the thermal diffuse scattering is given as

$$I^T(\theta) \propto f^2\left\{1-\exp\left[-2B(\sin\theta/\lambda)^2\right]\right\} \quad (3.37)$$

where f and exp are the atomic scattering factor and the Debye-Waller factor, respectively; and the parameter B for each element is available [38]. Based on Eq. 3.37, the intensity of thermal diffuse scattering is zero at the origin of the reciprocal space, increases with the scattering angle at the maximum value, and then decreases gradually. As mentioned in Section 3.1, energy loss due to phonon excitation is so small the effect of thermal diffuse scattering cannot be removed with the energy filter installed on a conventional electron microscope. Instead, the effect of thermal diffuse scattering can be suppressed by cooling a specimen with a cooling holder. On the other hand, the intensity of inelastically scattered electrons due to the plasmon excitation is theoretically given as

$$I^{p}(\Theta) \propto \frac{\Theta_p}{\Theta^2 + \Theta_p^2} \tag{3.38}$$

where Θ equals 2θ (θ is the diffraction angle) and Θ_p equals $\hbar\omega_p/(2E)$ with plasmon energy $\hbar\omega_p$ and the incident electron energy E [39]. Because the plasmon energy generally ranges from several electron volts to 30 eV and is much smaller than the incident electron energy (100–1000 keV), the background due to plasmon excitation has high intensity around the transmitted beam and decreases drastically with an increase in the scattering angle. Furthermore, the background is enhanced and spreads owing to the dynamic diffraction effect. Thus, for analysis of convergent beam electron diffraction (CBED) patterns and evaluation of the superlattice reflections and diffuse scattering that result from atomic ordering or atomic displacement (or both), removal of the background due to plasmon excitation is important. Examples of applying of energy filtering to a CBED pattern and diffuse scattering of an alloy semiconductor and a shape memory alloy are discussed below.

Evaluation of CBED Pattern. In addition to the transmitted beam, the inelastically scattered electrons form a strong background around the strong reflections due to the dynamical diffraction effect; the elastically scattered electrons should be eliminated to obtain an accurate intensity distribution of CBED patterns with little background noise. Figure 3.58 shows energy-filtered CBED patterns of hematite (α-Fe_2O_3). Sharp white and black bands in the CBED pattern are observed utilizing an omega-type energy filter. The electron diffraction pattern in Fig. 3.58b is observed by selecting the elastically scattered electrons with a slit of energy width 20 eV. As described in Section 5.1.2, the posi-

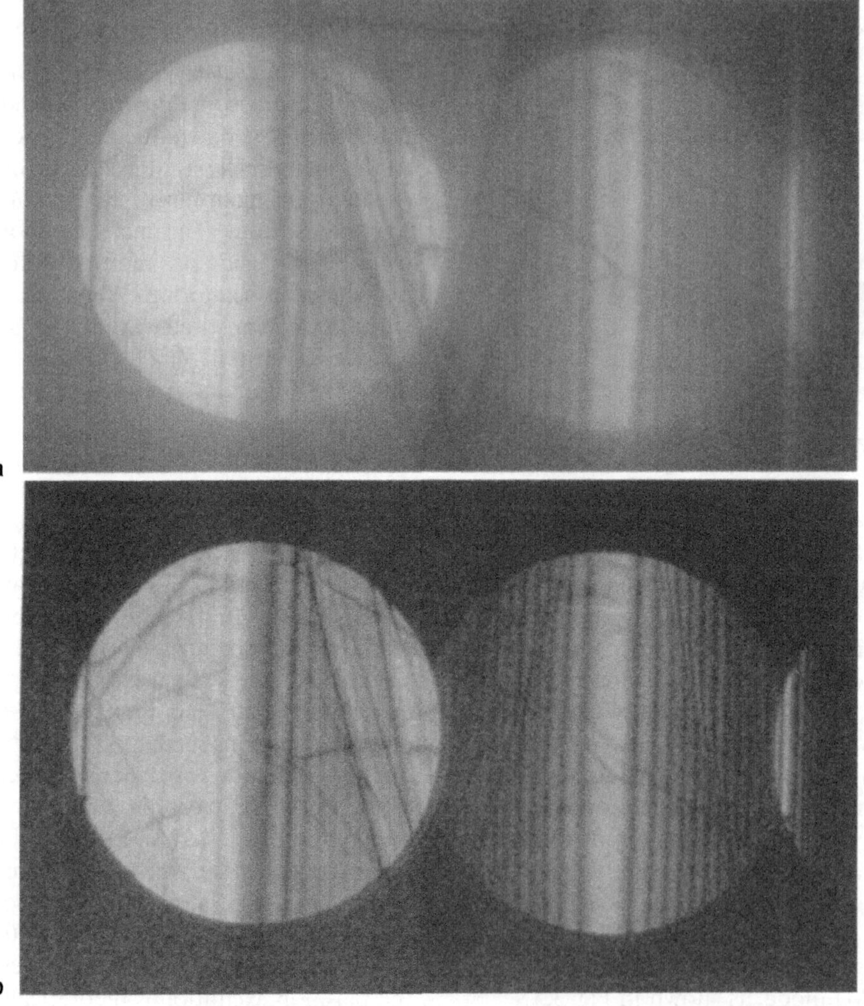

Fig. 3.58. CBED patterns of hematite (α-Fe_2O_3) without **a** and with **b** an energy filter

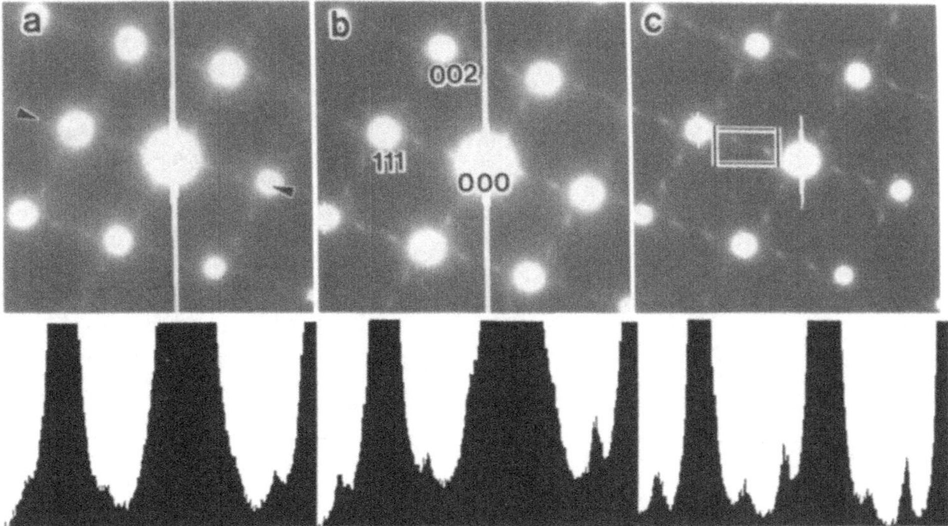

Fig. 3.59. **a** Electron diffraction patterns of $Al_{0.48}In_{0.52}As$. Intensity profiles between the *arrowheads* are shown at the bottom. **b** Electron diffraction pattern observed at 107K. **c** Electron diffraction pattern observed with the energy filter at 107K

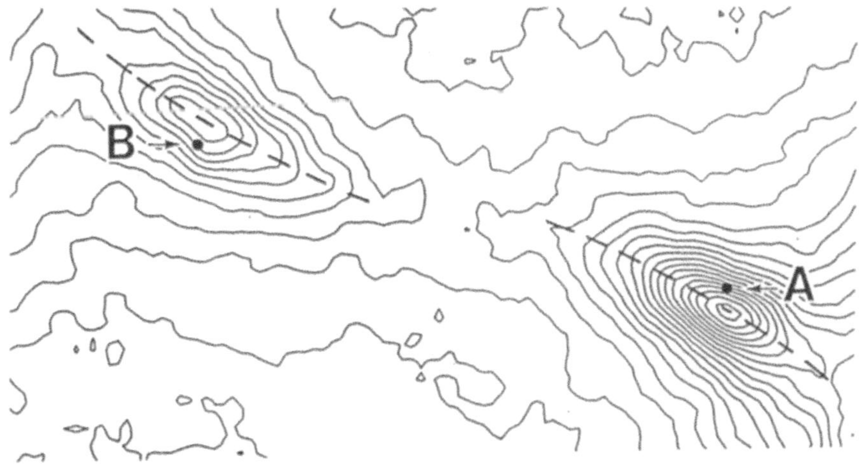

Fig. 3.60. Intensity profile of the rectangle region in Fig. 3.59c. *A* and *B*, positions of the ⅓ ⅓ ⅓ and ⅔ ⅔ ⅔ reflections

tions of intensity maxima and minima can be determined based on an evaluation of the intensity distribution in the bands, and thus the thickness of the crystal can be evaluated accurately. Quantitative information about crystal structure has also been obtained from energy-filtered CBED patterns [40].

Evaluation of Diffuse Scattering. Figure 3.59 shows electron diffraction patterns of $Al_{0.48}In_{0.52}As$, which attracts much attention in optical and electronic devices [41]. The patterns were observed with a CCD camera. Whereas Fig. 3.59a is a conventional electron diffraction pattern without an omega-type filter (unfiltered), Fig. 3.59b,c show the patterns observed at 107K without and with the filter. The width of the energy-selecting slit for Fig. 3.59c is set at 10 eV [42], and diffuse scattering resulting from the ordering of constituent elements Al and In is easily observed. Especially around the transmitted beam, the background caused by plasmon excitation is effectively removed, and the intensity distribution and peak position can be accurately determined, as shown in Fig. 3.60 [42]. In addition to analysis of the energy-filtered electron diffraction patterns, the processing of high-resolution electron microscope images has been utilized to

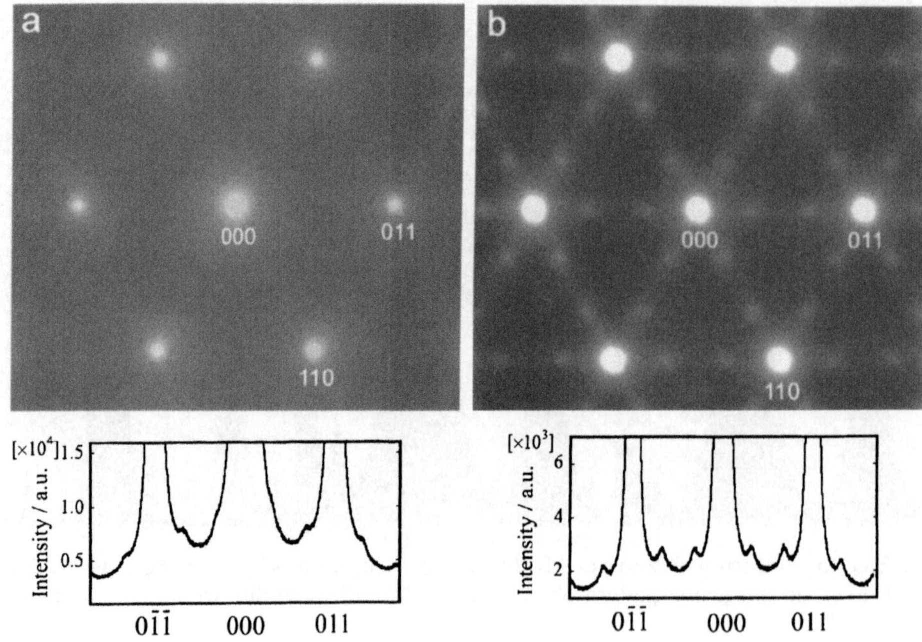

Fig. 3.61. Energy filtering on diffuse scattering of a shape memory alloy $Ti_{50}Ni_{48}Fe_2$. Electron diffraction patterns (*top*) and their intensity profiles (*bottom*) without and with the energy filter

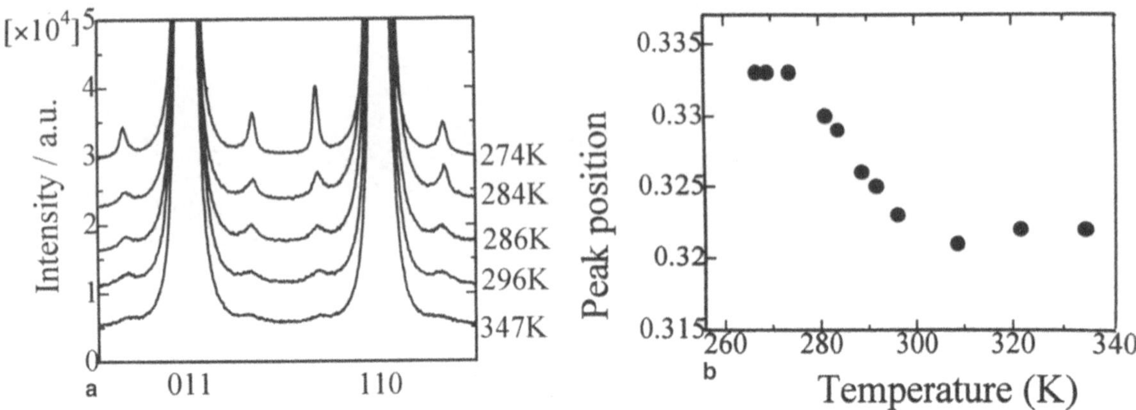

Fig. 3.62. Change in intensity profile **a** and peak position **b** of diffuse scattering as a function of temperature

understand the characteristic feature of the short-range ordered structure [43].

Figure 3.61 is another example of diffuse scattering in energy-filtered electron diffraction patterns observed with a shaped memory alloy $Ti_{50}Ni_{48}Fe_2$. Because of energy filtering with an energy-selecting slit of 20 eV in this case, the weak diffuse scattering at around one-third position between the fundamental reflections is clearly observed in Fig. 3.61b. Furthermore, the intensity profile and the peak position of the diffuse scattering have been quantitatively determined as a function of temperature, as shown in Fig. 3.62. The microstructure corresponding to the so-called precursor effect in the martensitic phase transformation, which had been a controversial problem, was clarified by energy-filtered electron diffraction and in situ dark-field electron microscopy [44,45].

References

1. Shuman H (1981) Parallel recording of electron energy loss spectra. Ultramicroscopy 6:163

2. Krivanek OL, Ahn CC, Keeney RB (1987) Parallel detection electron spectrometer using quadrupole lenses. Ultramicroscopy 22:103
3. Terauchi M, Kuzuo R, Satoh F, Tanaka M, Tsuno K, Ohyama J (1991) Performance of a new high-resolution electron energy-loss spectroscopy microscope. Microsc Microanal Microstruct 2:351
4. Lee C-W, Ikematsu Y, Shindo D (2002) Measurement of mean free paths for inelastic electron scattering of Si and SiO_2. J Electron Microsc, p 143
5. Oikawa T, Bando Y, Hosoi J, Kokubo Y (1985) Advantages of a HVEM in electron energy loss spectroscopy: in situ experiments with high voltage electron microscopes. In: Proceeding of the international symposium on "behavior of lattice imperfections in materials—in situ experiments with HVEM", Osaka, 409
6. Egerton RF (1978) Quantitative energy-loss spectroscopy. In: Johari O (ed) Scanning electron microscopy. SEM, Chicago, 1, p 133
7. Isaacson MS, Silcox J (1976) Report of a workshop on analytical electron microscopy held at Cornell University. Ultramicroscopy 2:89
8. Suenaga K,Ténce M, Mory C, Colliex C, Kato H, Okazaki T, Shinohara H, Hirahara K, Bandow S, Iijima S (2000) Element-selective single atom imaging. Science 290:2280
9. Egerton RF (1996) Electron energy-loss spectroscopy in the electron microscope, 2nd edn. Plenum, New York
10. Raether H (1980) Excitation of plasmons and interband transitions by electrons. In: Springer tracts in modern physics, **vol 88**. Springer, Berlin Heidelberg New York
11. Auerhammer JM, Rez P (1989) Dipole-fobidden excitations in electron-energy-loss spectroscopy. Phys Rev B40:2024
12. Jones W, Sparrow TG, Williams BG, Herley PJ (1984) Evidence for the formation of single crystals of sodium metal during the decomposition of sodium aluminum hydride: an electron microscopic study. Mater Lett 2:377
13. Kunz C (1966) Messung Charakteristischer Energieverluste von Elektronen an leichtoxydierbaren Metallen im Ultrahochvakuum. Z Phys 196:311
14. Kloos T (1973) Plasmaschwingungen in Al, Mg, Li, Na and K angeregt durch schnelle Elektronen. Z Phys 265:225
15. Daniels J, Festenberg CV, Raether H, Zeppenfeld K (1970) Optical constants of solids by electron spectroscopy. In: Springer tracts in modern physics, **vol 54**. Springer, Berlin Heidelberg New York, p 78
16. Zeppenfeld K, Raether H (1966) Energieverluste von 50keV-Elektronen an Ge und Si. Z Phys 193:471
17. Sueoka O (1965) Plasma oscillation of electrons in Be, Mg, Al, Si, Ge, Sn, Sb and Bi. J Phys Soc Jpn 20:2203
18. Aiyama T, Yada K (1974) Plasmon damping in Be, Mg, Al, Si, Ge and Sn. J Phys Soc Jpn 36:1554
19. Oikawa T, Hosoi J, Inoue M, Harada Y (1982) Applications of electron energy analyzer. JEOL News 20E:8
20. Nishino D, Nakafuji A, Yang J-M, Shindo D (1998) Precise morphology analysis on platelet-type hematite particles by transmission electron microscopy. ISIJ Int 38:1369
21. Egerton RF (1980) Instrumentation and software for energy-loss microanalysis. In: Johari O (ed) Scanning electron microscopy. SEM, Chicago, 1, p 41
22. Lee Y-S, Murakami Y, Shindo D, Oikawa T (2000) Effect of scattering angle on energy loss near-edge structure of h-BN. Mater Trans JIM 41:555
23. Ichinose H, Zhang Y, Ishida Y, Ito K, Nakanose M (1996) Morphology, atomic structure and electron structure of artificial diamond grain boundary. JEOL News 32E:16
24. Hosoi J, Oikawa T, Kokubo Y (1985) Computed deconvolution in electron energy loss spectroscopy (EELS). J Electron Microsc 34:73
25. Shindo D, Oh-ishi K, Hiraga K, Syono Y, Hojou K, Furuno S (1991) Oxygen K-edge fine structure of $La_{2-x}M_xCuO_{4-y}$ (M = Sr, Ba and Ca) studied by electron energy loss spectroscopy. Mater Trans JIM 32:872
26. Murakami Y, Shindo D, Chiba H, Kikuchi M, Syono Y (1999) Charge ordering in $Bi_{1-x}Ca_xMnO_3$ ($x \geq 0.75$) studied by electron-energy-loss spectroscopy. Phys Rev B59:6395
27. Pearson DH, Fultz B, Ahn CC (1988) Measurements of 3d state occupancy in transition metals using electron energy loss spectrometry. Appl Phys Lett 53:1405
28. Murakami Y, Shindo D, Otsuka K, Oikawa T (1998) Electronic structure changes associated with a martensitic transformation in $Ti_{50}Ni_{48}Fe_2$ alloy studied by electron energy-loss spectroscopy. J Electron Microsc 47:301
29. Shindo D, Hiraga K, Tsai A-P, Chiba A (1993) Cu $L_{2,3}$ white lines of Cu compounds studied by electron energy loss spectroscopy. J Electron Microsc 42:48
30. Zanchi G, Perez JP, Sevely J (1975) Adaptation of a magnetic filtering device on a one megavolt electron microscope. Optik 43:495
31. Castaing R, Hennequin JF, Henry L, Slodgian G (1967) The magnetic prism as an optical system. In: Septier A (ed) Focusing of charged particle, Academic Press, New York, p 265
32. Krivanek OL, Gubbens AJ, Dellby N (1991) Developments in EELS instrumentation for spectroscopy and imaging. Microsc Microanal Microstrct 2:315
33. Hashimoto H, Makita Y, Nagaoka N (1992) Atomic structure images formed by core loss electrons.

In: Bailey GW, Bentley J, Small JA (eds) Proceedings of the 50th Annual EMSA Meeting, Boston, p 1194
34. Ajika N, Hashimoto H, Endo H, Yamaguchi K, Tomita M, Egerton RF (1983) Construction of analyzer for energy-filtered lattice image. In: Proceedings of the Japanese Society of Electron Microscopy, annual meeting, p 134 (in Japanese)
35. Taya S, Taniguchi Y, Nakazawa E, Usukura J (1996) Development of γ-type energy filtering TEM. J Electron Microsc 45:307
36. Oikawa T, Sasaki H, Matsuo T, Kokubo Y (1982) Elemental filtergrams obtained by means of electron energy analyzer combined with image storage system. In: Bailey GW (ed) Proceedings of the 40th annual EMSA meeting, Washington, DC, p 736
37. Segawa M, Taniyama A, Shindo D (1998) HREM observation of the interface between Laves-phases and matrix phases in Inconel 718 by using a high-voltage electron microscope. ISIJ Int 38:1375
38. Sears VF, Shelley SA (1991) Debye-Waller factor for elemental crystals. Acta Cryst A47:441
39. Ferrel RA (1956) Angular dependence of the characteristic energy loss of electrons passing through metal foils. Phys Rev 101:554
40. Spence JCH, Zuo JM (1992) Electron microdiffraction. Plenum, New York
41. Gomyo A, Makita K, Hino I, Suzuki T (1994) Observation of a new ordered phase in $Al_xIn_{1-x}As$ alloy and relation between ordering structure and surface reconstruction during molecular-beam-epitaxial growth. Phys Rev Lett 72:673
42. Shindo D, Spence JCH, Gomyo A (1995) Evaluation of electron diffuse scattering by energy filtering. In: Shin KS, Yoon JK, Kim SJ (eds) Proceedings of the 2nd Pacific Rim International Conference on advanced materials and processing. Korean Institute of Metals and Materials, Seoul, p 1077
43. Shindo D, Gomyo A, Zuo J-M, Spence JCH (1996) Short-range ordered structure of $Ga_{0.47}In_{0.53}As$ studied by energy-filtered electron diffraction and HREM. J Electron Microsc 45:99
44. Murakami Y, Shindo D (1999) Lattice modulation preceding to the R-phase transformation in a $Ti_{50}Ni_{48}Fe_2$ alloy studied by TEM with energy-filtering. Mater Trans JIM 40:1092
45. Shindo D, Murakami Y (2000) Advanced transmission electron microscopy study on premartensitic state of $Ti_{50}Ni_{48}Fe_2$. Sci Technol Adv Mater 1:117

4. Energy Dispersive X-ray Spectroscopy

A typical analytical electron microscopic method (i.e., energy dispersive X-ray spectroscopy, or EDS, sometimes called EDX or EDXS) is described in this chapter. Although some improvement in the resolution of EDS has been attempted, there has been no significant modification introduced in the practice and application of EDS in comparison with electron energy-loss spectroscopy (EELS). Still, this method is the most standard and reliable one in the field of analytical electron microscopy and is widely used.

The analytical technique and quantitative methods are discussed based on the principles of emission of characteristic X-rays and their detectors. In the latter part, ALCHEMI, which is performed using the electron channeling effect for locating additives or impurities, and bremsstrahlung emission, which is observed in crystalline specimens, are explained in detail, with the presentation of typical experimental data.

4.1 Emission of Characteristic X-rays

As noted in Section 1.2.1, the emission of X-rays is a phenomenon produced by inner-shell excitation with incident electrons. That is, when an inner shell electron transits to a high-energy level, the hole in the inner shell is filled by an electron at a high energy level, resulting in emission of a characteristic X-ray with energy between those two energy levels. When the electron in the higher-energy level transits to the lower-energy level, only the transition, which corresponds to the difference of the quantum number of orbital angular momentum $\Delta l = \pm 1$ due to the selection rule may occur. The emission process of the Kα_1 characteristic X-ray due to $L_3 \rightarrow K$ transition resulting from the hole formation in the 1s inner shell (K shell) is shown in Fig. 1.6, although various characteristic X-rays can be emitted owing to hole formation in the K and L shells. As indicated in Fig. 4.1, the species of characteristic X-rays are given by symbols such as Kα_1, Kα_2, and so on. The "K" indicates the X-ray emitted when the electron transits to the K shell from the outer shell. In the same way, L and M series correspond to the X-ray emitted when the electron transits to the L shell and the M shell, respectively. The Greek letter and the number (e.g., α_1 and α_2) specifies the species of the transition corresponding to the specific characteristic X-ray. These symbols were first introduced by Siegbahn.[1] When the energy difference between the characteristic X-rays (e.g., Kα_1 and Kα_2) is too small to distinguish, they are represented by K$\alpha_{1,2}$ or Kα.

Because characteristic X-rays have specific energy corresponding to each element, the element can be identified from the peak energy; and the content of the element in the compound can be analyzed from the integrated intensity of the peak. It should be noted that when the excited state of the atom with a hole changes to the ground state, an Auger electron may be emitted instead of the characteristic X-ray emission. In general, the emission probability of characteristic X-rays increases with the increase in atomic number, whereas the emission probability of Auger electrons decreases complementarily. Thus, EDS is more useful for heavy elements, especially when the content of the element is small.

---

[1] In addition to symbols such as Kα_1 and Kα_2 proposed by Siegbahn, others such as K-L_3 and K-L_2, which specify the transition between the shells represented in the symbols, are used in IUPAC representations [1,2].

Quantum Numbers Specifying the Energy Levels of Atomic Electrons

According to the Pauli principle, the electron state is specified by four quantum numbers (n, l, m_l, m_s), and any two electrons can never have the same four numbers in an atom. When an electron is moving along a circular orbit around the nucleus with the charge Ze, the energy of the electron is given as

$$E_n = \frac{2\pi^2 m_0 Z^2 e^4}{h^2} \cdot \frac{1}{n^2} \tag{4.1}$$

where m_0 is the mass of the electron, and h is Plank's constant. The positive number n, which determines this energy, is called the principal quantum number. When the elliptical orbit is considered, taking into account the deviation from the circular orbit, the quantum number of orbital angular momenta (l) related to the major and minor axes of the ellipse should be introduced. For specifying the details of the electron state further, the magnetic quantum number (m_l), indicating the effect of the magnetic field on the angular momentum of the electron, and the spin quantum number (m_s), indicating the effect of the spin motion of the electron on the total angular momentum, should be introduced. The shell indicated by the principal quantum number n is represented by the commonly used symbol, and the maximum number of the electrons in the shell is given by $2n^2$ (Table 4.1). The shell specified by the principal number in Table 4.1 is sep-

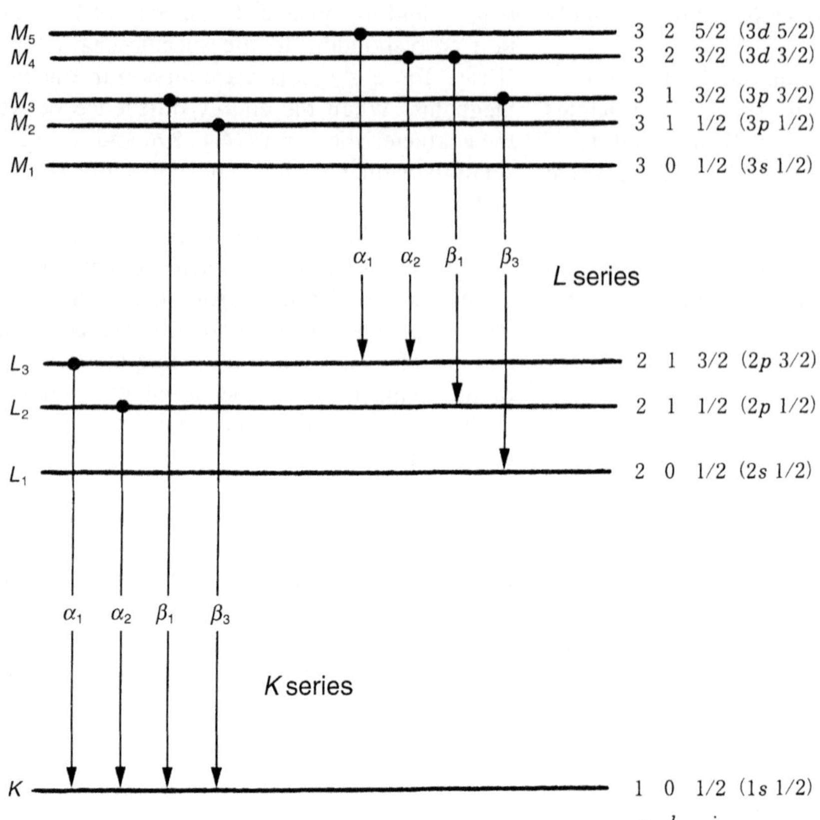

Fig. 4.1. Energy levels of atomic electrons and species of characteristic X-rays

Table 4.1. Symbol of the shell corresponding to the principal quantum number and maximum number of electrons in the shell.

Principal quantum number (n)	1	2	3	4	5	6	7	...
Symbol of the shell	*K*	*L*	*M*	*N*	*O*	*P*	*Q*	...
Maximum number of electrons in the shell ($2n^2$)	2	8	18	32	50	72	98	...

arated into subshells with the quantum number of orbital angular momentum l ($l = 0,1,2 \ldots < n - 1$). The subshells are also represented by the symbol, and the maximum number of electrons in the subshell is given by $2(2l + 1)$ (Table 4.2). The symbol of the subshell means sharp (s), principal (p), diffuse (d), fundamental (f), and so on. Also, the magnetic quantum number m_l (integer) and spin quantum number m_s assume values given by $-l < m_l < l$, $m_s = \pm 1/2$.

When the interaction of the spin momentum and the orbital momentum are considered (spin-orbital interaction), the subshell is specified by the principal number n, the quantum number of orbital angular momentum l, and total angular quantum number j. The latter (j) takes values obtained by $|l - 1/2| < j < l + 1/2$; thus it takes two values for $l > 0$. If the subshell (l) has branches due to j, the maximum number of electrons in the j subshell is $(2j + 1)$, which means that the number of electrons in the subshell $j = l + 1/2$ is $(2j + 1) = 2(l + 1)$ and that in the subshell $j = l - 1/2$ is $(2j + 1) = 2l$. The species of the subshells are specified by these three quantum numbers, such as $2s$ 1/2 (see the left of Fig. 4.1). As shown in Fig. 4.1 (left), the subshells are also represented with the symbol of the principal quantum number with a lower-case figure (sometimes Roman lower-case figure), and the edges of the inner-shell electron excitation in EELS presented in Section 3 are specified by this symbol.

Table 4.2. Symbol of the subshell corresponding to the quantum number of orbital angular momentum and the maximum number of electrons in the subshell.

Quantum number of orbital angular momentums (l)	0	1	2	3	4	5	6	...
Symbol of the subshell	*s*	*p*	*d*	*f*	*g*	*h*	*i*	...
Maximum number of electrons in the subshell ($2(2l + 1)$)	2	6	10	14	18	22	26	...

4.2 X-ray Detectors and Their Principles

Characteristic X-rays emitted from a specimen can be analyzed with EDS or wavelength dispersive X-ray spectroscopy (WDS). EDS is utilized for transmission electron microscopy owing to its high detection efficiency. X-rays emitted from the specimen enter the detector through the collimator. Figure 4.2 shows the cross section of an EDS detector installed at a specimen chamber. Figure 4.3 diagrams the EDS system. For an EDS detector, a solid-state detector (SSD) of a high-purity Si single crystal doped with a small amount of Li is generally used. The SSD is thought to be a solid ionization chamber where electric charges proportional to the energy of the incident X-ray are produced. The charges accumulate at the field effect transistor (FET), and the pulse voltage with the pulse height value proportional to the charges is then produced. The pulse voltage is analyzed by a multichannel pulse height analyzer, and the pulse number corresponding to each pulse height is displayed. In this way, the spectrum with a horizontal axis of X-ray energy and a vertical axis of the photon number is obtained. To stabilize Li in Si and suppress the thermal noise, the EDS detector is usually cooled with liquid N_2. Two kinds of detector window protect the detector, and their properties and treatment are different from each other.

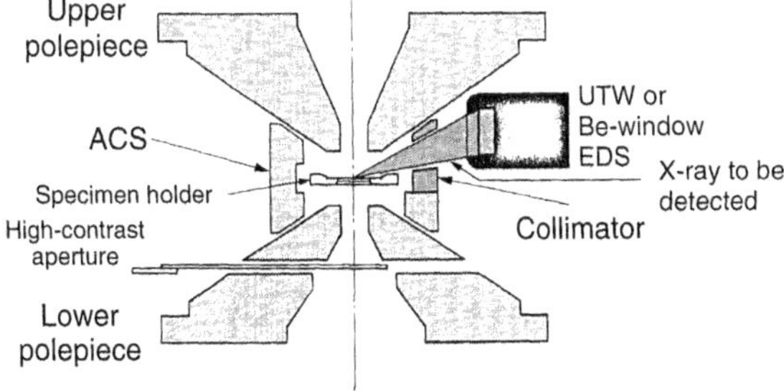

Fig. 4.2. Cross section of an EDS detector installed on a specimen chamber. *ACS*, anticontamination system; *UTW*, ultra-thin window

4.2.1 Beryllium Window Type

The vacuum condition inside the detector is maintained with beryllium film of 8–10 μm thickness, and its treatment is rather easy. However, because of the absorption of low-energy X-rays in Be film, elements lighter than Na (Z = 11) cannot be analyzed.

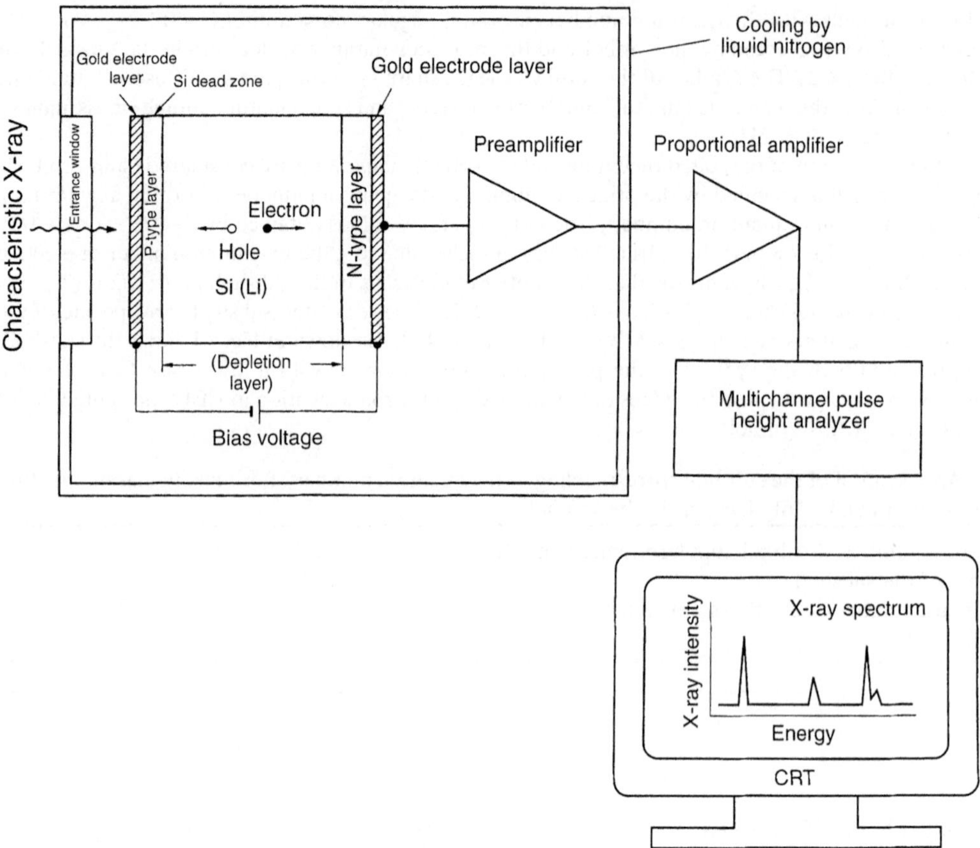

Fig. 4.3. Constitution of EDS system

4.2.2 Ultra-thin Window Type

Light elements up to C (Z = 6) can be analyzed, as the protector film (detector window) consists of Al-doped thin organic film 0.3–0.5 μm in thickness. On the other hand, the vacuum condition inside should be carefully maintained. Because the light elements can be analyzed, this detector is mostly installed on transmission electron microscopes these days.

Moreover, an EDS detector of the windowless type without a window film is developed, and light elements up to B (Z = 5) can be analyzed. However, this type of detector is used for scanning electron microscopy (SEM) with low accelerating voltages to reduce damage to the detector caused by back-scattered electrons.

Resolution of the EDS is about 150 eV, which is definitely worse than EELS and WDS. In WDS where crystals for spectroscopy such as LiF (lattice constant d_{200} = 0.2013 nm) and RAP (rubidium acid phthalate, d_{001} = 1.305 nm), are utilized for the analysis. Light elements up to B (Z = 5) can usually be analyzed, and the resolution is around 10 eV. However, WDS has low detection efficiency, so the acquisition time is rather long. Thus, EDS with high detection efficiency is installed on transmission electron microscopes (TEMs) for detecting weak-intensity X-rays emitted from small areas of thin film, and WDS is installed on electron probe microanalyzers (EPMA) for analyzing strong X-ray intensity emitted from a bulk specimen. Figure 4.4 shows an energy dispersive X-ray spectrum obtained from the intermetallic compound $Ni_{70}Al_{25}Fe_5$. For comparison, part of a wave dispersive X-ray spectrum obtained from the same specimen with WDS of an EPMA is shown in Fig. 4.5. The single peak of an Al-K line observed with EDS in Fig. 4.4 is clearly split into two lines (Kα and Kβ) in Fig. 4.5. Moreover, the Fe Kβ line partly overlaps the Ni Kβ with EDS, whereas there is no overlapping of these lines with WDS. Improving the resolution with EDS was attempted recently, and the resolution comparable to that of EELS was obtained with the microcalorimetry method [3,4].

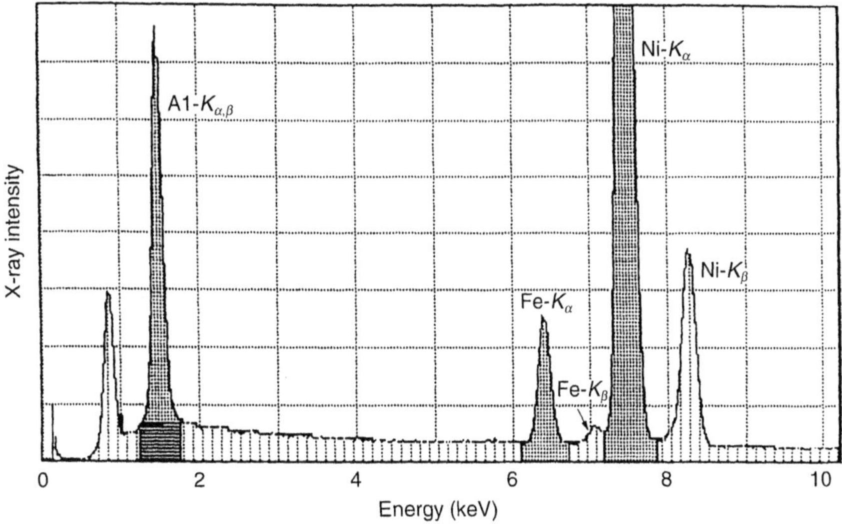

Fig. 4.4. Energy dispersive X-ray spectrum obtained from intermetallic compound $Ni_{70}Al_{25}Fe_5$

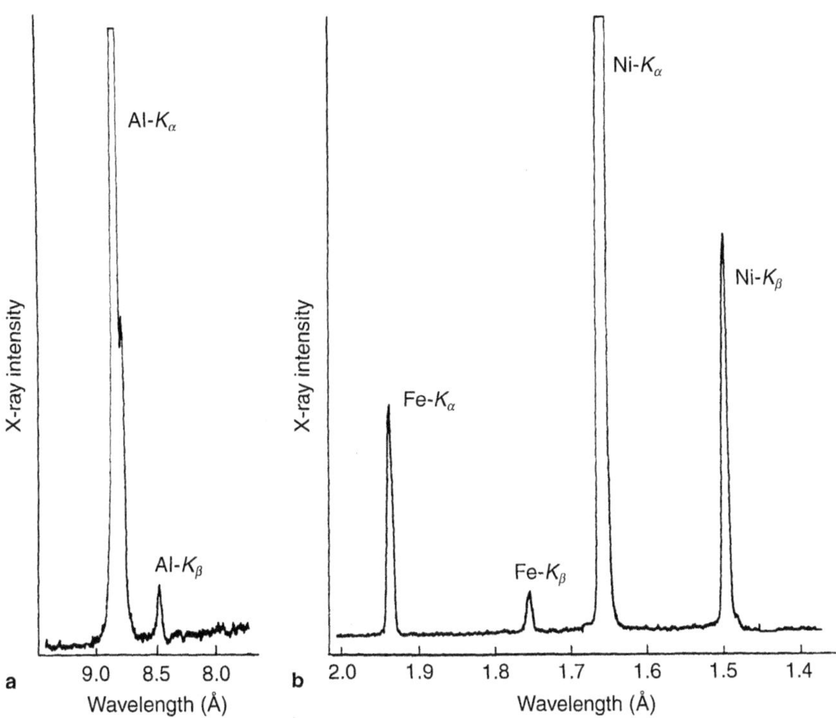

Fig. 4.5. Wavelength dispersive X-ray spectrum obtained from $Ni_{70}Al_{25}Fe_5$

4.3 Analytical Technique for EDS

4.3.1 Detection of X-rays

Note that continuous X-rays and scattered X-rays from the specimen holder may enter the X-ray detector and cause background, with occasional production of artifacts in an X-ray spectrum. A beryllium specimen holder should be used to reduce the background. Also, a specimen-supporting grid that does not contain the constituent elements of the specimen should be chosen. When the intensity of the X-rays emitted

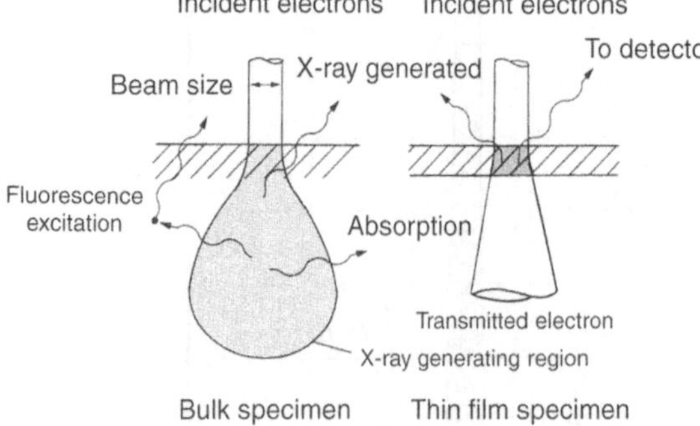

Fig. 4.6. Broadening of the incident electron beam in a specimen

from the specimen becomes high owing to strong electron intensity, the counting loss increases. The magnitude of the counting error is called dead time (T_{dead}) and is given by

$$T_{dead} = (1 - R_{OUT}/R_{IN}) \times 100(\%) \qquad (4.2)$$

where R_{IN} and R_{OUT} are counting fractions for the input and output, respectively. When the dead time is more than 50%, the intensity of the incident electrons should be decreased to optimize the counting efficiency.

4.3.2 Spatial Resolution

Figure 4.6 shows spreading of the incident electron beam in the area where X-rays will be produced. Most of the incident electrons can transmit the thin film, which is prepared for TEM. Thus, diffusion of incident electrons in a thin specimen is much less than for a bulk specimen, as shown in Fig. 4.6 (left), and the spatial resolution in the analysis is expected to be high. Hence, with analytical electron microscopy diffusion of the incident electrons in the specimen affects the spatial resolution. The main factors that determine spatial resolution are accelerating voltage, beam diameter, specimen thickness, and specimen density.

4.3.3 Peak/Background Ratio

As with EELS (see Sect. 3.3.1), the characteristic X-ray/background intensity ratio is called P/B, and a high P/B ratio should be expected for quantitative analysis [5]. According to Zaluzec, the intensity of the characteristic X-rays detected is given by

$$N = (I\sigma\omega p N_0 \rho C t \Omega \varepsilon)/(4\pi M) \qquad (4.3)$$

where I is the incident electron intensity; C is the content (wt%); σ is the ionization cross section; t is the specimen thickness; ω is the fluorescent yield; Ω is the detection solid angle; p is the production fraction of characteristic X-rays to be analyzed; ε is the detection efficiency; N_0 is Avogadro's number; M is the atomic number; and ρ is the density.

When the accelerating voltage increases, the intensity of the characteristic X-ray slightly decreases, but the background X-ray decreases largely resulting in an increase in the P/B ratio generally.

4.3.4 Elemental Mapping Method

The analysis where the incident electron beam is stopped at a point in a specimen and the X-ray from the area is detected is called the *point analysis*. The electron beam can be scanned on the specimen with a beam-scanning system, and the specific X-ray intensity is measured. When the brightness signal corresponding to the characteristic X-ray intensity measured is displayed on the CRT by synchronizing it with the position signal, a two-dimensional X-ray intensity can be obtained. This observation mode is called the *elemental mapping method* and is effective for analyzing the distribution of the constituent element in two dimensions.

Figure 4.7 shows the X-ray spectrum and the elemental mapping images of the so-called SIALON. The signal intensity of the elemental mapping image corresponds to the net elemental signal, which is obtained by subtracting the background. Compared with the energy-loss spectrum shown in Fig. 3.24, the X-ray energy dispersive spectrum has a low background and can be used for quantitative analysis. By utilizing a microprobe with the field emission gun (FEG), elemental mapping images with a resolution better than 1 nm can be obtained.

Figure 4.8 shows an elemental mapping image of an Sm-Co magnet [6]. It is well known that this material forms a cell structure consisting of two phases. One is the matrix phase of Sm_2Co_{17} and is called the 2:17 phase; the other is the cell boundary phase of $SmCo_5$, called the 1:5 phase. When elemental mapping images are obtained with an FEG, the so-called Z-phase, which contains a lot of Zr, is observed as planar precipitates about 1 nm in thickness. Also, from the elemental mapping

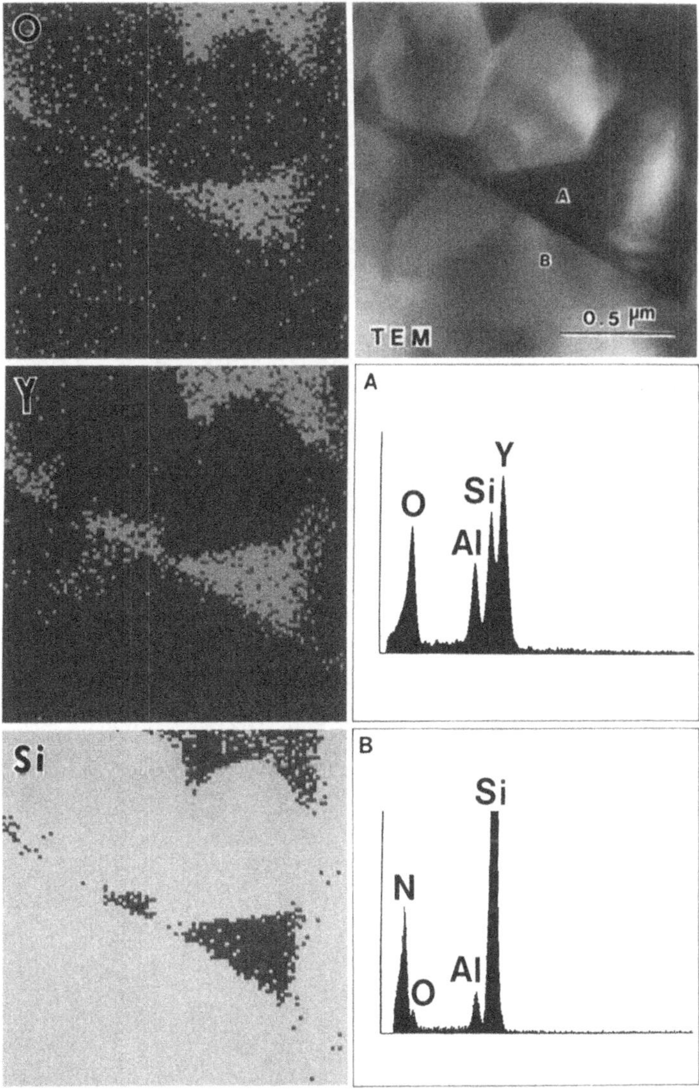

Fig. 4.7. Elemental mapping images of SIALON and its X-ray spectra obtained from areas *A* and *B*

image, the distributions of Cu and Fe (which largely affect the magnetic property of the material) can be visualized. It is especially noted that the 1:5 phase contains much Cu in the magnet (Fig. 4.8b) annealed at the optimum condition. By comparing Zr and Fe mapping images, it is seen that the content of Fe is lower at the Z-phase.

4.3.5 Detection Limit of EDS

The detection limit has been discussed with the minimum mass fraction (MMF) (see ref. [7], for example). It has been determined by the peak intensity (or P/B) in a spectrum and the acquisition time. Recently, the performance of electron microscopes has been much improved by utilizing an FEG and a detector of high sensitivity.

Watanabe and Williams [8] reported an example of atomic order detection. Two atoms of Mn in a thin foil of Cu (0.12 wt%) were detected by a 300 kV scanning electron microscope with an FEG. Also, P atoms of 2 atomic percentage (at%) in O-N-O dielectric layers in a semiconductor device has been reported by Kawasaki et al. [9]. In that study a 0.9-nm diameter probe was used for a specimen thickness of 50 nm using a 200 kV electron microscope with an FEG; this corresponds to the detection of about twenty atoms. A minimum detectable concentration in EDS with a standard specimen has been roughly estimated to be about 1 at%.

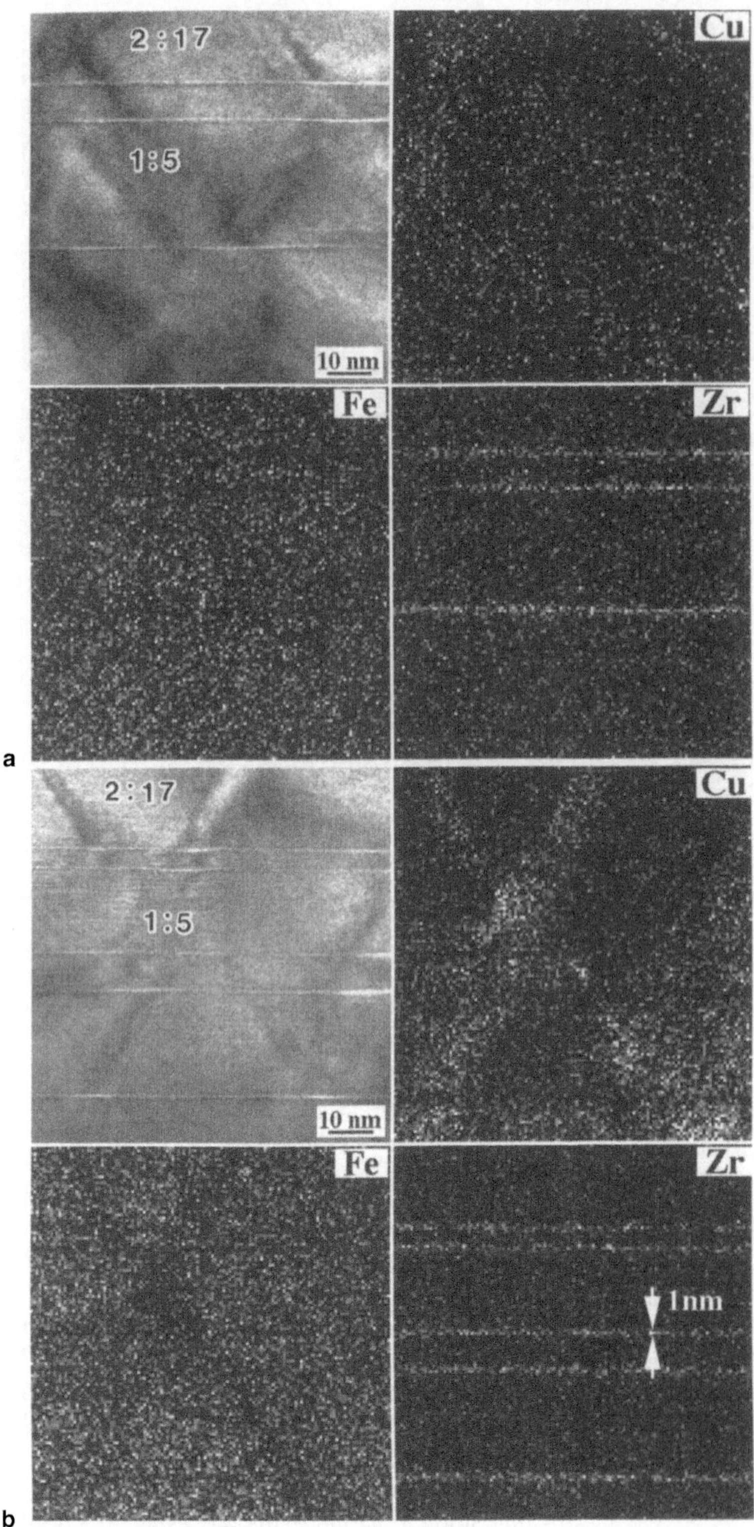

Fig. 4.8. Elemental mapping images of an Sm-Co magnet. **a** Isothermal aging. **b** Stepped aging. **a,b** The *top left* images are STEM images

4.4 Quantitative Analysis

4.4.1 k-Factor

When a specimen is thin (thin film approximation), the characteristic X-ray of element A detected can be derived from Eq. 4.3 as

$$N_A = (I\sigma_A\,\omega_A\,p_A\,N_0\,\rho C_A\,t\Omega\varepsilon_A)/(4\pi M_A) \quad (4.4)$$

Thus, the ratio of the characteristic X-rays from elements A and B in compound A-B is given by

$$N_A/N_B = (\sigma_A\,\omega_A\,p_A\,C_A\,\varepsilon_A\,M_B)/(\sigma_B\,\omega_B\,p_B\,C_B\,\varepsilon_B\,M_A) \quad (4.5)$$

The ratio of the contents (wt%) of elements A and B (C_A/C_B) is given with the intensity ratio of X-rays N_A/N_B, as

$$C_A/C_B = [(\sigma_B\,\omega_B\,p_B\,\varepsilon_B\,M_A)/(\sigma_A\,\omega_A\,p_A\,\varepsilon_A\,M_B)](N_A/N_B) = k_{AB}(N_A/N_B) \quad (4.6)$$

where k_{AB} is called the k-factor, or the Cliff-Lorimer factor [10], and is given by

$$k_{AB} = (\sigma_B\,\omega_B\,p_B\,\varepsilon_B\,M_A)/(\sigma_A\,\omega_A\,p_A\,\varepsilon_A\,M_B) \quad (4.7)$$

From Eq. 4.6 it is seen that the ratio of the constituent element contents is determined by X-ray intensities and the k-factor. The ratio of atomic contents is similarly given by

$$C'_A/C'_B = [(\sigma_B\,\omega_B\,p_B\,\varepsilon_B)/(\sigma_A\,\omega_A\,p_A\,\varepsilon_A)](N_A/N_B) = k'_{AB}(N_A/N_B) \quad (4.8)$$

indicating that the ratio is determined by the factor k'_{AB}.

The so-called ZAF correction should be carried out for compositional analysis on a bulk specimen with an EPMA. That is, one should take into account the difference in electron scattering due to the atomic number (Z), the effect of absorption (A) of the X-ray, and the change in fluorescence (F) yield.[1]

4.4.2 Practice in Quantitative Analysis

As the composition of a compound can be given through Eq. 4.6, it is seen that the accuracy of the compositional analysis depends on the precision of determining the k-factor. There are two ways to determine the k-factor: One is calculation based on the theoretical formulation, and the other is experimental determination with a standard specimen of a known composition.

For theoretical calculation of the k-factor, the following equation is used for the ionization cross section.

$$\sigma = \frac{6.51\times 10^{-20}}{E_c^2 U^{d_s}}\cdot n_s b_s \ln(c_s U) \quad (4.9)$$

where n_s indicates the number of electrons in the shell involved (i.e., 2, 8, and 18 for the K, L, and M shells) [11]; U is the ratio of the incident electron energy E to the ionization energy E_c, being called overvoltage; and b_s, c_s, and d_s are parameters [12] obtained for the K, L, and M shells [5,11,13,14]. Usually each EDS system on a TEM is equipped with some software containing the k-factor calculated for various elements. By using this software, the composition is easily determined according to Eq. 4.6. Also, for overlapping characteristic X-rays, the intensity distribution of each peak can be calculated with a reference peak of the characteristic X-ray in the software; and thus the quantitative analysis is carried out. In general, however, the accuracy of the analysis with k-factor theoretically calculated is low. Quantitative compositional analysis with high accuracy is not expected, especially in compounds containing elements whose atomic numbers are largely different.

To calculate the k-factor experimentally, it is useful to prepare a standard compound with a known composition that is expected to be close to that of the specimen to be analyzed. If the standard specimen is prepared and the k-factor is determined experimentally with Eq. 4.6, errors in compositional determination of less than a few percent are expected. When there is overlapping of characteristic X-rays, the intensity distribution in each peak is obtained using the spectrum of the standard specimen.

Examples of compositional analyses are shown in Table 4.3. Those numbered (1) and (2) are compositions determined with the theoretically calculated k-factor, and (3) is the result obtained with the standard specimen. All the data are compared

[1] This means that the fluorescence of the characteristic X-ray with the energy E_1 is strongly enhanced, with the characteristic X-ray of energy E_2 being slightly larger than E_1:

$$E_2 = E_1 + \Delta E$$

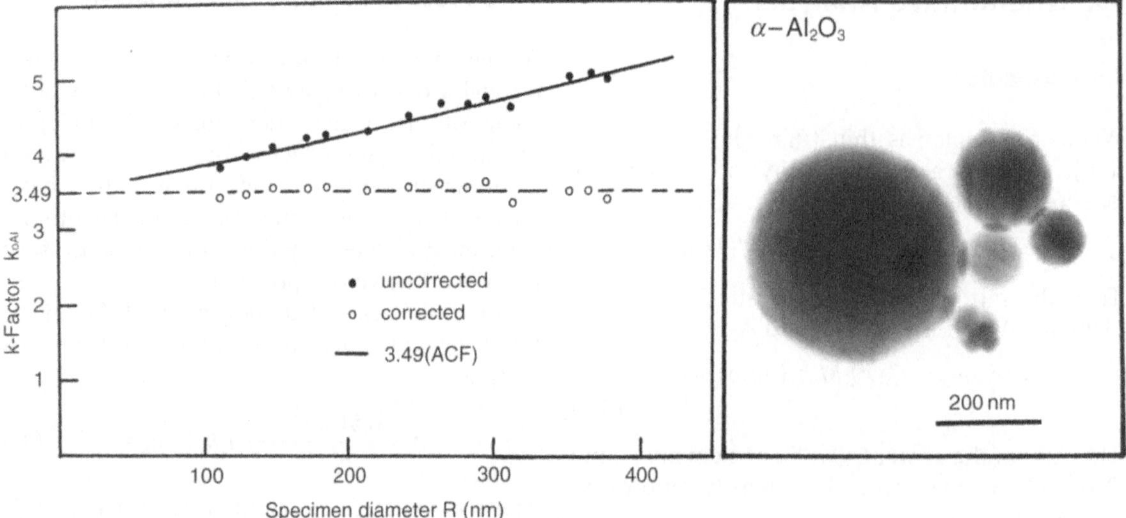

Fig. 4.9. Thickness dependence of the k-factor for oxygen and Al (k_{OAl}) on thickness. *Filled circles*, observations; *open circles*, values corrected taking into account the absorption effect on X-rays. Specimens of spherical alumina (α-Al_2O_3) particles are shown on the *right*

Table 4.3. Result of compositional analysis by EDS compared with EPMA.

	EDS	EPMA
(1) Al-Pd-Cr	$Al_{73.0}Pd_{10.3}Cr_{16.7}$	$Al_{77.3}Pd_{9.0}Cr_{13.7}$
(2) Al-Pd-Co	$Al_{66.8}Pd_{11.2}Co_{22.0}$	$Al_{74.6}Pd_{9.0}Co_{16.4}$
(3) Al-Pd-Mn	$Al_{69.8}Pd_{13.1}Mn_{17.1}$	$Al_{69.2}Pd_{13.3}Mn_{17.5}$

Compositions (1) and (2) are determined with the k-factor theoretically calculated; composition (3) is the result obtained with the standard specimen
EDS, energy dispersive X-ray spectroscopy; EPMA, electron probe microanalysis

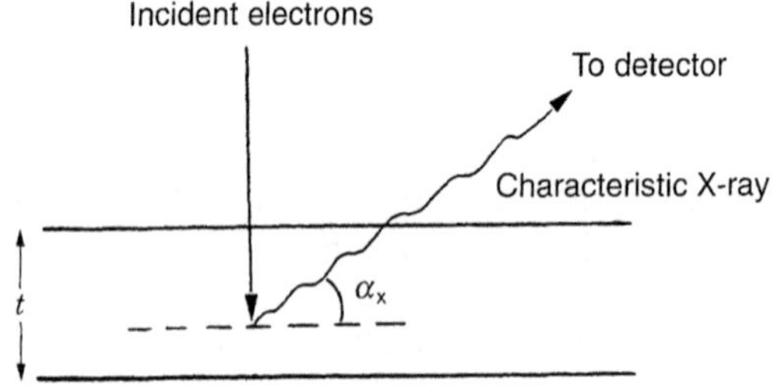

Fig. 4.10. Path length of X-rays generated in a specimen of a parallel-plate shape

with the results obtained by EPMA, which generally provides more accurate compositions than TEM-EDS. It is seen that the accuracy of composition with the calculated k-factor is low, as the atomic numbers of Al and Pd are largely different.

4.5 Notice of Quantitative Analysis

4.5.1 Absorption of X-rays by a Specimen

During quantitative analysis of the composition shown in Section 4.4 the neglect of X-ray absorption is assumed. Figure 4.9 shows the thickness dependence of the k-factor for oxygen and Al (k_{OAl}) on thickness. Because the specimens are spherical alumina (α-Al_2O_3) particles, thickness is easily estimated from their spherical shape. With increasing thickness, the k-factor obtained by neglecting X-ray absorption deviates from the value obtained for extremely thin crystals or the values obtained taking into account the absorption. Note that deviation is prominent for X-rays with largely different energies.

When the specimen has the parallel-plate shape shown in Fig. 4.10, correction for the absorption effect is carried out by evaluating the path length of the X-rays. The intensity of the characteristic X-ray from element A is given with the mass absorption coefficient.

$$N_A = \frac{I\sigma_A\omega_A p_A N_0 \rho C_A t\Omega\varepsilon}{4\pi M_A} \cdot \int_0^t \exp\left[-\left(\frac{\mu}{\rho}\right)_A \cdot \operatorname{cosec}\alpha_x \cdot \rho z\right] dz$$
$$= \frac{I\sigma_A\omega_A p_A N_0 \rho C_A t\Omega\varepsilon}{4\pi M_A} \cdot \frac{1-\exp\left[-\left(\frac{\mu}{\rho}\right)_A \cdot \operatorname{cosec}\alpha_x \cdot \rho t\right]}{\left(\frac{\mu}{\rho}\right)_A \cdot \operatorname{cosec}\alpha_x \cdot \rho} \quad (4.10)$$

Thus, the equation for the ratio of the contents of elements A and B is modified with the correction factor *CF*.

$$C_A/C_B = (k_{AB} \cdot CF) \cdot (N_A/N_B) \quad (4.11)$$

where

$$CF = \frac{\left(\frac{\mu}{\rho}\right)_A}{\left(\frac{\mu}{\rho}\right)_B} \cdot \frac{1-\exp\left[-\left(\frac{\mu}{\rho}\right)_B \cdot \operatorname{cosec}a_x \cdot \rho t\right]}{1-\exp\left[-\left(\frac{\mu}{\rho}\right)_A \cdot \operatorname{cosec}a_x \cdot \rho t\right]} \quad (4.12)$$

The method for determining the k-factor by extrapolation is also proposed on the basis of the values obtained at different thickness regions [15].

4.5.2 Statistical Error

When the compound contains a small number of elements, a long acquisition time is necessary to reduce the statistical error. If the X-ray peak has a Gaussian distribution (normal distribution), the standard deviation σ is given with the integrated intensity of the characteristic X-ray.

$$\sigma = N^{1/2} \quad (4.13)$$

For example, the error for $\Delta N = 3\sigma$ given the 99.7% reliability can be evaluated as

$$\text{Error} = (\pm 3N^{1/2}/N) \times 100$$
$$= \pm 3N^{-1/2} \times 100\,(\%) \quad (4.14)$$

From the above equation, the error is evaluated to be about ±1% for $N = 10^5$.

4.5.3 Escape Peak, Sum Peak, and Others

4.5.3.1 Escape Peak

When the characteristic X-ray with energy E_x enters the Si(Li) detector, electric charges proportional to the energy are produced, as noted in Section 4.2. When the Si of the detector, being in the excited state, emits the Kα characteristic X-ray (1.740 keV) and the characteristic X-ray escapes from the detector, the artificial X-ray peak with energy

$$E_{esc} = E_X - 1.740\ \text{keV} \quad (4.15)$$

appears in the spectrum. This peak is called the *escape peak*. Discriminating this peak from the standard characteristic X-ray peaks is important for identifying the constituent elements of a compound.

4.5.3.2 Sum Peak

When the two X-rays enter the detector almost at the same time incidentally, these X-rays cannot be distinguished, so the artificial peak appears at the energy being equal to the sum of the the energies of the two X-rays. The peak is called the *sum peak*. If there are strong X-rays with energies E_1 and E_2, the sum peaks may appear at energies of $E_1 + E_2$ as well as at twice the energies of E_1 and E_2. The sum peak often appears when the dead time of the detector is greatly increased (see Section 4.3.1).

4.5.3.3 System X-rays and Spurious X-rays

X-rays emitted from the illumination lens system of an electron microscope are called *system X-rays*. X-rays emitted from areas other than the area being investigated with electron beam illumination are called spurious X-rays. To check the effect of these X-rays, it is useful to evaluate them by illuminating the vacuum region near the specimen edge with the incident electron beam.

4.5.4 Crystalline Specimens

4.5.4.1 Coherent Bremsstrahlung

As noted in Section 3.1, when an electron goes near the atomic nucleus, the direction of the incident electron changes owing to the attraction force resulting in the emission of X-rays (i.e., *bremsstrahlung*). For axial illumination, the electron passes along the atomic row and produces bremsstrahlung periodically, resulting in enhanced X-ray intensity. Eventually, bremsstrahlung forms small peaks in the spectrum. This phenomenon is called *coherent bremsstrahlung*. Observations and theoretical interpretations were carried out by Überall [16] in 1956 and by Barbiellini et al. [17] in 1962. Following these studies, coherent bremsstrahlung has been observed and analyzed by Spence et al. with analytical electron microscopy [18–20].

The characteristic features in coherent bremsstrahlung can be summarized as follows.

1. The energy of the X-ray peak due to coherent bremsstrahlung is inversely proportional to the lattice spacing (L) in the direction parallel to the electron beam.
2. The energy of the X-ray peak also strongly depends on the direction of the incident electron beam, the accelerating voltage, and the detector angle (α) (Fig. 4.11).
3. The X-ray peak due to coherent bremsstrahlung appears at a low-energy region in the spectrum.

Figure 4.11 shows the geometrical configuration of the incident electron beam, crystal plane, and a detector. The energy of the X-ray peak due to coherent bremsstrahlung E_{CB} is given by

$$E_{CB} = \frac{hc\beta_r}{L[1-\beta_r \cos(90+\alpha_L)]} \tag{4.16}$$

where h and β_r are Planck's constant and the velocity ratio (v/c) of the electron and the light, respectively; L corresponds to the lattice spacing (in Å); and α_L is the angle between the lattice plane and the detector. X-ray spectra observed in Mn-Zn ferrite with [100] incidence are shown in Fig. 4.12 [21]. The encircled numbers indicate X-ray peaks due to coherent bremsstrahlung corresponding to the nth order Laue zone. With an increase in the accelerating voltage from 100 kV to 300 kV, the X-ray peak due to coherent bremsstrahlung shifts to a higher energy. Table 4.4

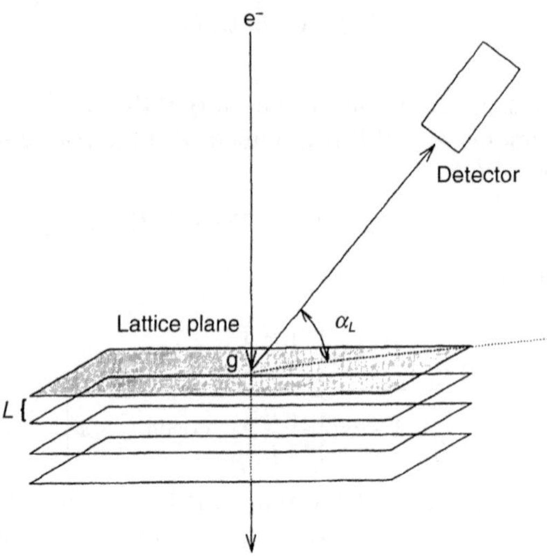

Fig. 4.11. Geometrical configuration of the incident electron beam, crystal plane, and a detector

Table 4.4. Peak energy of coherent bremsstrahlung in Mn-Zn ferrite with [100] incidence.

Accelerating voltage (kV)	[100] β (= v/c)	Lattice spacing L(A)	Laue zone (n)	E_{CB}(keV) Calculated	Experimental
300	0.7765	4.240	②	1.320	1.32
		2.827	③	1.980	1.99
		2.120	④	2.641	2.65
		1.696	⑤	3.301	3.30
250	0.7410	4.240	②	1.125	1.29
		2.827	③	1.927	1.92
		2.120	④	2.569	2.56
		1.696	⑤	3.211	3.21
200	0.6953	4.240	②	1.236	1.23
		2.827	③	1.854	1.86
		2.120	④	2.473	2.46
		1.696	⑤	3.091	3.09
150	0.6343	4.240	②	1.168	—
		2.827	③	1.752	(1.75)
		2.120	④	2.336	2.33
		1.696	⑤	2.920	2.93
100	0.5482	4.240	②	1.063	—
		2.827	③	1.594	1.58
		2.120	④	2.126	2.11
		1.696	⑤	2.657	2.65

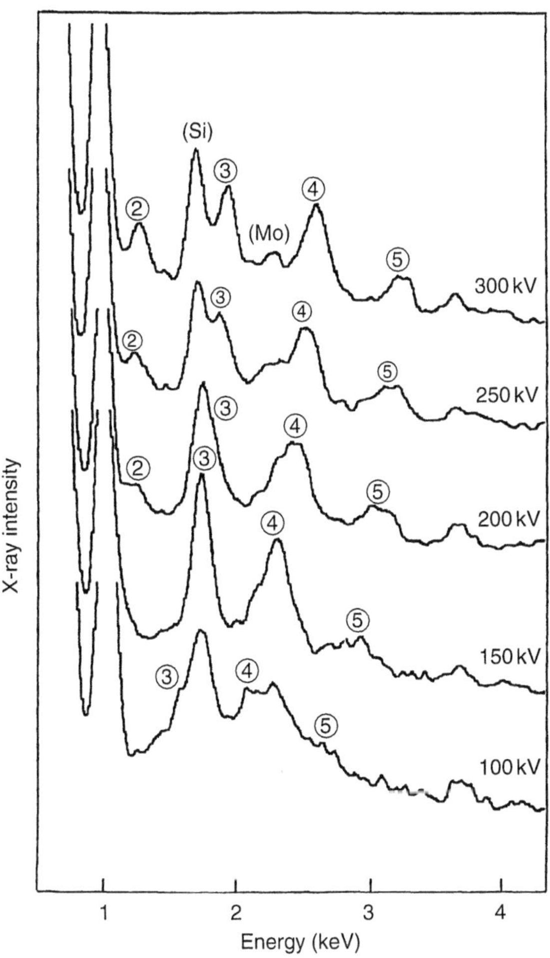

Fig. 4.12. X-ray spectra observed in Mn-Zn ferrite with [100] incidence

compares the observed energies and calculated energies obtained with Eq. 4.16.

Figure 4.13 shows X-ray spectra observed with [211] incidence and tilted illumination by 10° from the [211] axis. As can be seen in Fig. 4.13, the X-ray peak due to coherent bremsstrahlung is not observed by tilting the crystal by more than several degrees from the axial incidence condition.

4.5.4.2 Effect of Electron Channeling

In crystalline specimens, in addition to the X-ray peak due to coherent bremsstrahlung, the intensity distribution of characteristic X-rays is modified by the diffraction effect or channeling effect. Figure 4.14 shows X-ray spectra of superconducting oxide $Tl_2B_2CuO_y$ observed under two diffraction conditions. The two spectra are normalized with the intensity of the Ba-L line. Figure 4.14a shows the so-called nonchanneling condition, where the reflections with low indices are not strongly excited. Figure 14.4b corresponds to one of the channeling conditions; that is, systematic reflections along the c* axis are excited. Because of the strong diffraction effect, a decrease in the characteristic X-ray peak from Tl is clearly noted. Figure 4.15 shows X-ray spectra observed in an Al-Fe-Cu quasicrystal. The spectra are normalized with the intensity of the Al-K line. The intensities of characteristic X-rays from Fe and Cu observed in axial incidences parallel to the fivefold and twofold axes are largely different from those observed under nonchanneling conditions [22].

As demonstrated in Figs. 4.14 and 4.15, the diffraction effect influences the X-ray spectrum and makes the accuracy of the compositional analysis low. Thus, after observing electron diffraction patterns or lattice images with the axial illumination condition, one should note the diffraction effect on the compositional analysis. Especially when the structure and composition at the interface is investigated with EDS coupled with high-resolution electron microscopy, one should compare the X-ray spectra obtained at the matrix phase under axial illumination (channeling condition) and the nonchanneling condition. If there is a difference between the spectra, special attention should be paid to a comparison of the compositions between the matrix and the interface.

For determining the composition, the diffraction effect or the channeling effect should be avoided. However, this channeling effect can be utilized positively to locate the occupation site of additives and impurities as explained below.

4.6 ALCHEMI

4.6.1 Principle

It has long been known that the incident electron beam is localized in a crystal due to the diffraction effect. The first attempt to determine the atomic position by utilizing this effect was proposed by Cowley in 1964 [23]. The experiment for locating the impurities was carried out by Batterman with X-rays in 1969 [24]. It was found that the channeling effect is useful for distinguishing the substitutional type from the interstitial type and furthermore for specifying the location of the impurity atoms. It is necessary to prepare a large

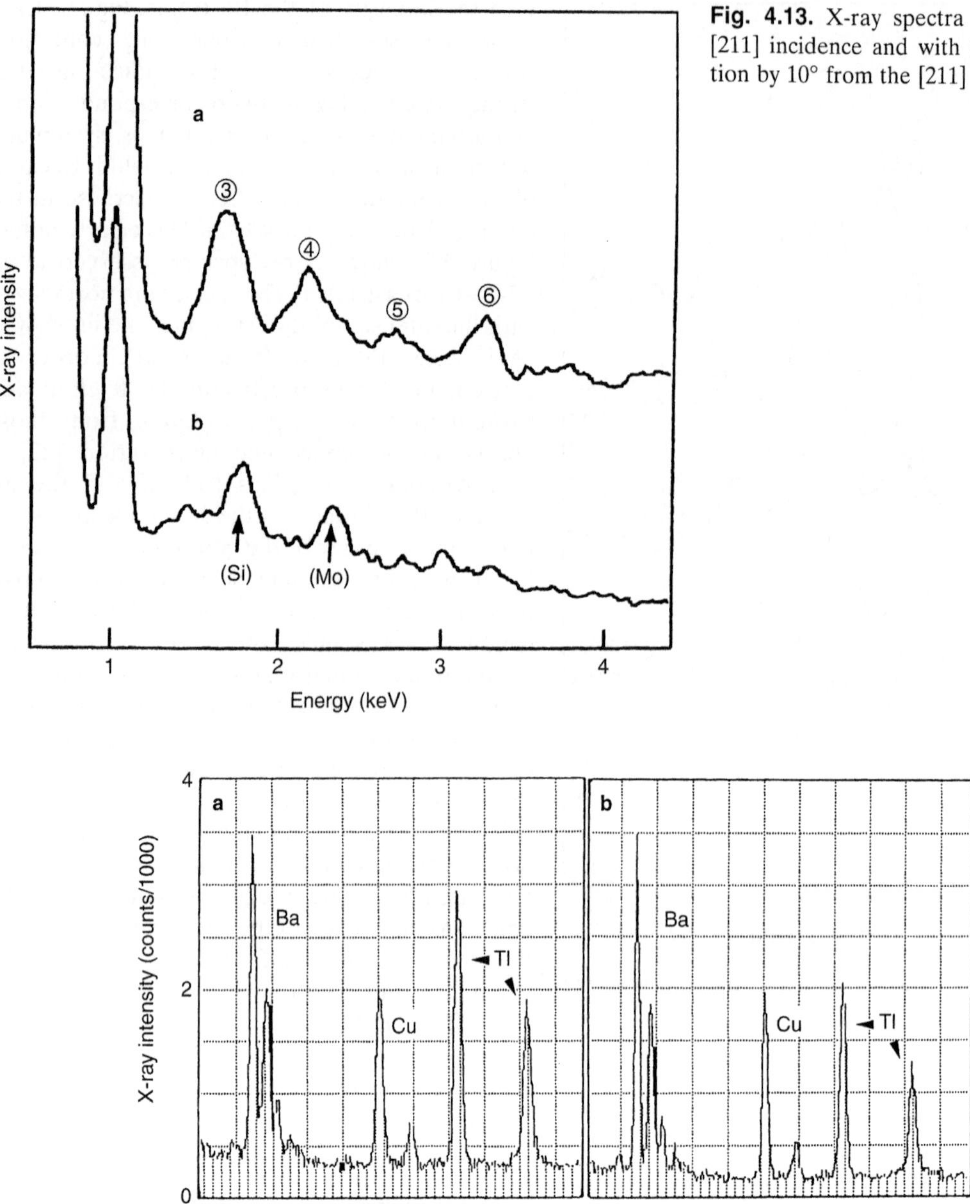

Fig. 4.13. X-ray spectra observed with [211] incidence and with tilted illumination by 10° from the [211] axis

Fig. 4.14. X-ray spectra of superconducting oxide $Tl_2B_2CuO_y$ under two diffraction conditions: nonchanneling **a** and channeling **b**

single crystal, however, and thus the application is limited to a few elements such as Si. ALCHEMI (*a*tom *l*ocation *ch*anneling *e*nhanced *mi*croanalysis) [25] has adapted the principle to compounds with the electron beam to locate a small amount of additives or impurities of the substitutional type. In this case, the diffraction condition can easily be controlled by observing electron diffraction patterns with an analytical electron microscope installed with an EDS system; thus, the technique can be applied to a small single crystal of less than 1 μm. Also, the technique does not require knowledge of the incident electron intensity distribution in a crystal and the k-factor, which is necessary for quantitative compositional analysis; the site of the impurities can be quantitatively

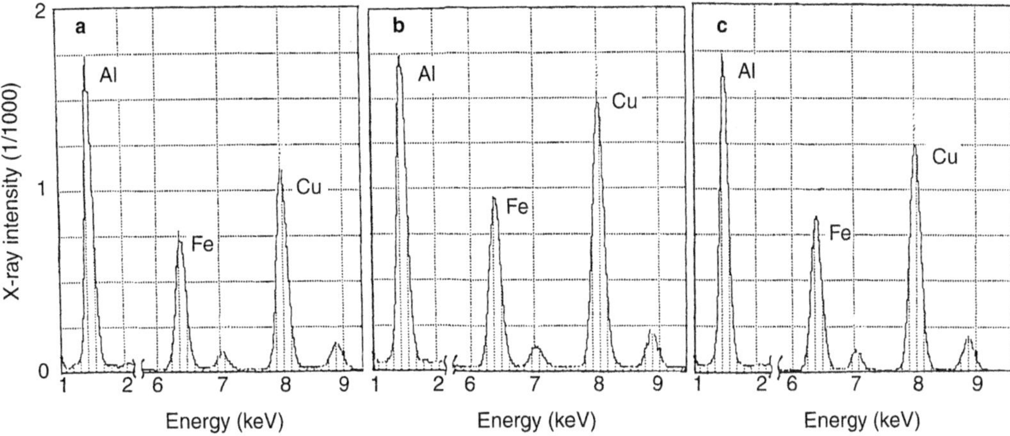

Fig. 4.15. X-ray spectra observed from an Al-Fe-Cu quasicrystal. **a** Nonchanneling condition. **b**,**c** Axial incidences parallel to the five-fold and two-fold symmetry axes, respectively

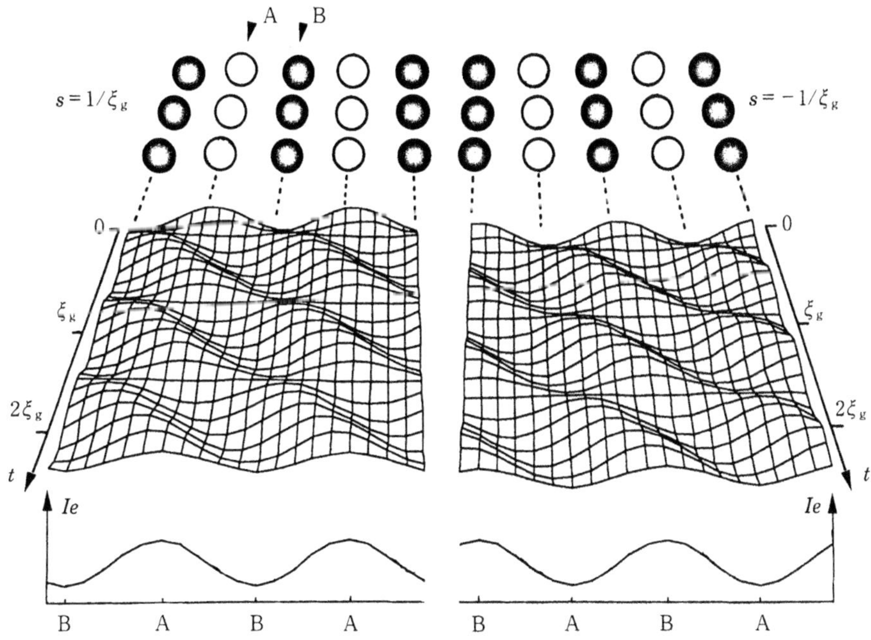

Fig. 4.16. Density of the incident electrons in compound A-B as a function of crystal thickness. Density of the incident electron averaged along the thickness are shown below for $s = 1/\xi_g$ and $s = -1/\xi_g$

determined from only the change of the characteristic X-ray intensity. Thus, the technique has been widely applied to locating additives and impurities in various compounds.

To understand the principle of ALCHEMI, consider the density of the incident electron in a compound consisting of a light element A and a heavy element B, as shown in Fig. 4.16. Here we set the crystal so that one superlattice reflection is excited and the electron beam is incident almost parallel to the atomic planes alternately consisting of A and B atoms. For the two-beam condition (i.e., $s = 0$), two waves form whose densities have maxima at B atomic planes (Bloch wave 1) and at A atomic planes (Bloch wave 2); and the sum of the electron wave densities becomes constant in the crystal. Here s is the deviation parameter, and it is defined as having a positive value when the reflec-

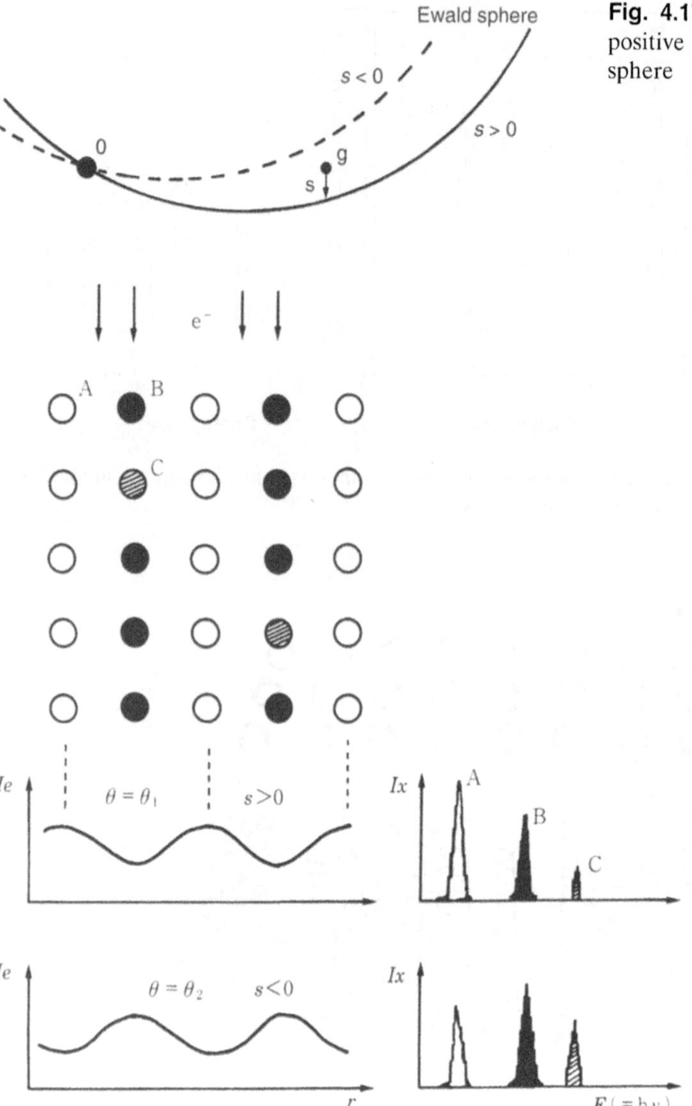

Fig. 4.17. Illustration showing the sign of s. s is positive when the reflection is inside the Ewald sphere

Fig. 4.18. Principles of ALCHEMI

tion is inside the Ewald sphere, as shown in Fig. 4.17. If the crystal is tilted somewhat from the two-beam condition, the sum of the electron wave density at the A atomic planes becomes different from that at the B atomic planes. The change in electron density as a function of thickness and the averaged density along the thickness are illustrated in Fig. 4.16 for $s = 1/\xi_g$ and $s = -1/\xi_g$ (where ξ_g is the extinction distance). In general, averaged electron density has maxima at light atomic planes (A) for $s > 0$, and at heavy atomic planes (B) for $s < 0$. In this way, because of the change in incident electron density, the characteristic X-ray of the light element is enhanced for $s > 0$, and that of the heavy element is enhanced for $s < 0$. When the electron density increases along the atomic planes as seen in Fig. 4.16 or also along the atomic columns, the phenomenon is called *electron channeling*. When channeling is induced along the atomic planes, it is called *planar channeling*; it is called *axial channeling* if the channeling is induced along the atomic columns.

Now, consider the case where a small amount of impurity C is included in compound A-B. By comparing the X-ray spectra observed for $s > 0$ and $s < 0$, the substitutional site of C can be detemined. For example, in Fig. 4.18 the intensity of characteristic X-rays from C increases for $s < 0$, similar to the B atoms; and it is concluded that C atoms take the site of B atoms.

A typical example is the result for a Ti-Al compound with an $L1_0$-type structure [26,27]. The

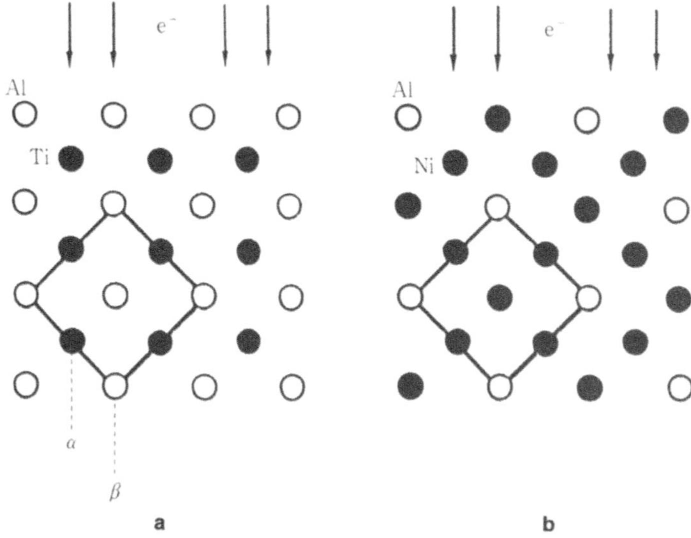

Fig. 4.19. Atomic arrangements of TiAl with the $L1_0$-type structure **a** and Ni_3Al with the $L1_2$-type structure **b**

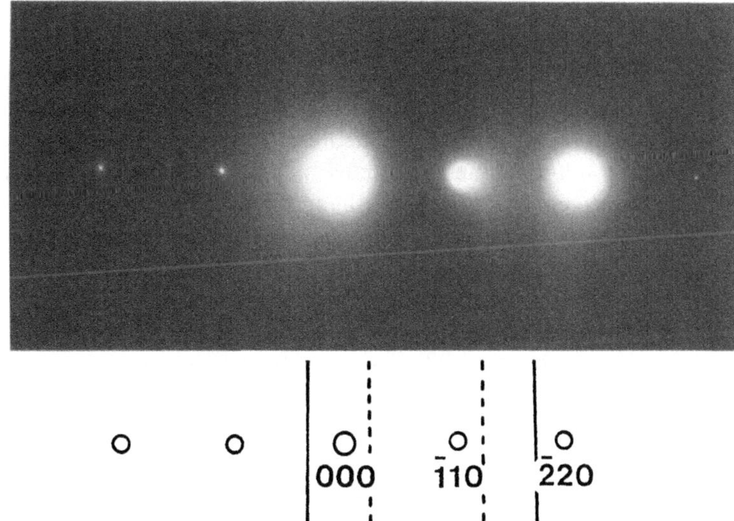

Fig. 4.20. Electron diffraction pattern at the condition ($s > 0$) of the channeling-enhanced microanalysis. Diffraction spots and Kikuchi lines are schematically indicated below. Two types of Kikuchi lines are indicated by *dotted lines* and *full lines* whose separation corresponds to $1/d_{\bar{1}10}$ and $1/d_{\bar{2}20}$, respectively

atomic arrangement of this structure projected along the [001] direction is shown in Fig. 4.19a, where there are atomic planes consisting alternately of Ti atoms (α-plane) and Al atoms (β-plane). The difference in the channeling effect from Ni_3Al with an $L1_2$-type structure is noted below in Section. 4.6.3. X-ray spectra were obtained near the two-beam condition with the $\bar{1}10$ reflection strongly excited (Fig. 4.20). X-ray spectra obtained from $Ti_{43}Al_{55}Nb_2$ under two diffraction conditions are shown in Fig. 4.21. The two spectra are normalized with the Al-K line, and it is clear that the characteristic X-ray from the heavy element (Ti) increases drastically for $s < 0$ as noted above. Because characteristic X-rays from Nb increase for $s < 0$, similar to Ti, it is concluded that Nb takes the Ti site. Quantitative evaluation of site occupancy is discussed in the next section.

4.6.2 Determination of Site Occupancy by ALCHEMI

Several methods for determining the site occupancy with ALCHEMI have been proposed. In their original paper, Spence and Tafto derived the formulation for the site occupancy by assuming that the amount of the impurities is small [25]. The formulation can be generalized taking into account the antisite occupancy of the host ele-

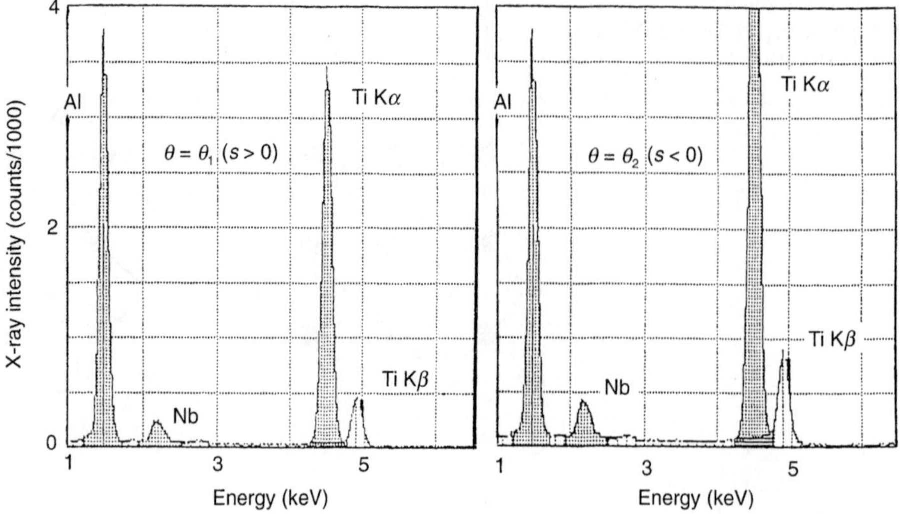

Fig. 4.21. X-ray spectra obtained from $Ti_{43}Al_{55}Nb_2$ under two diffraction conditions

ments for the case of nonstochiometric composition and a large amount of impurities or additives [26]. Although the site occupancy can basically be determined with the data obtained using two diffraction conditions, these methods utilized three diffraction conditions (one nonchanneling condition and two channeling conditions) to increase the accuracy of the analysis. Horita et al. derived the general formulation for determining site occupancy with the data obtained under the two diffraction conditions [28]. On the other hand, Rossouw et al. proposed the statistical methods for high accuracy by introducing the fitting parameters with many experimental data [29].

The method for determining site occupancy with three diffraction conditions is now explained. Consider the case where there is a small amount of additive C in compound A-B. The intensities of characteristic X-rays $N_A^{(n)}$, $N_B^{(n)}$, and $N_C^{(n)}$ from elements A, B, and C under three diffraction conditions n ($n = 1{:}s > 0, n = 2{:}s < 0, n = 3$: nonchanneling) are given as

$$N_A^{(n)} = P_A\left(A_\alpha I_\alpha^{(n)} + A_\beta I_\beta^{(n)}\right) \tag{4.17}$$

$$N_B^{(n)} = P_B\left(B_\alpha I_\alpha^{(n)} + B_\beta I_\beta^{(n)}\right) \tag{4.18}$$

$$N_C^{(n)} = P_C\left(C_\alpha I_\alpha^{(n)} + C_\beta I_\beta^{(n)}\right) \tag{4.19}$$

where $I_\alpha^{(n)}$ and $I_\beta^{(n)}$ are the incident electron densities on the α-plane (A atomic plane) and the β-plane (B atomic plane) averaged along the incident beam direction; P_A, P_B, and P_C are parameters that depend on the fluorescent yield and so on (see Eq. 4.3); A_α and A_β correspond to the atomic ratios of element A at the α- and β-planes, respectively; and $B_{\alpha,\beta}$ (or $C_{\alpha,\beta}$) is the atomic ratio of element B (or C) on the α- and β-planes, respectively. When the composition of the compound is given as $A_xB_yC_z$ ($x + y + z = 1$) and the fraction of the C atom at the α-plane is k, the atomic ratio of element C at the α-plane is given as

$$C_\alpha = z \cdot k \tag{4.20}$$

and the atomic ratio of C in the β-plane is given by

$$C_\beta = z(1-k) \tag{4.21}$$

Thus, $A_{\alpha,\beta}$, $B_{\alpha,\beta}$, and $C_{\alpha,\beta}$ are given with the composition x, y, z, and k. Based on these equations $I_\alpha^{(n)}$, $I_\beta^{(n)}$, P_A, P_B, and P_C can be eliminated, and the fraction of the C atom at plane k is given with the characteristic X-ray intensities $N_A^{(n)}$, $N_B^{(n)}$, and $N_C^{(n)}$ as

$$k = \frac{R_\alpha \cdot R_{\beta\alpha}^{(1)} - N_C \cdot R_{\beta\alpha}^{(2)}}{N_C\left(1-R_{\beta\alpha}^{(2)}\right) - R_\alpha\left(1-R_{\beta\alpha}^{(1)}\right)} \tag{4.22}$$

Here R_α, $R_{\beta\alpha}^{(n)}$, and N_C are

$$R_\alpha = \frac{N_A^{(1)}}{N_A^{(2)}} \cdot \frac{A_\alpha + A_\beta \cdot R_{\beta\alpha}^{(2)}}{A_\alpha + A_\beta \cdot R_{\beta\alpha}^{(1)}} \tag{4.23}$$

$$R_{\beta\alpha}^{(n)} = \frac{A_\alpha \cdot P_{AB} - B_\alpha \cdot \dfrac{N_A^{(n)}}{N_B^{(n)}}}{B_\beta \cdot \dfrac{N_A^{(n)}}{N_B^{(n)}} - A_\beta \cdot P_{AB}} \tag{4.24}$$

$$N_C = \frac{N_C^{(1)}}{N_C^{(2)}} \quad (4.25)$$

The derivation of these equations is in the Appendix at the end of this chapter. It is concluded that with knowledge of the basic crystal structure and composition, $A_{\alpha,\beta}$, $B_{\alpha,\beta}$, and $C_{\alpha,\beta}$ are given with the equation containing k; and k can be evaluated from the characteristic X-ray intensities $N_A^{(n)}$, $N_B^{(n)}$, and $N_C^{(n)}$.

For example, in the case of $Ti_{43}Al_{55}Nb_2$ shown in Fig. 4.21, the atomic ratios of Ti, Al, and Nb in the α- and β-planes are given as shown in Table 4.5. The fraction of Nb at the Ti site was determined to be 0.96 with the characteristic X-ray intensities in Table 4.6 [26].

The precision of the site occupancy can be evaluated from the statistical error. The site fraction k of the impurity is determined with the characteristic X-ray intensity $N_i^{(n)}$ of the constituent element i. Thus the statistical error due to the characteristic X-ray intensity $N_C^{(1)}$ of additive C under diffraction condition (1) is given by

$$\Delta k_C^{(1)} = k\left(N_A^{(1)}, N_B^{(1)}, N_C^{(1)} + \Delta N_C^{(1)}, N_A^{(2)} \ldots\right) \\ k\left(N_A^{(1)}, N_B^{(1)}, N_C^{(1)}, N_A^{(2)} \ldots\right) \quad (4.26)$$

where $\Delta N_C^{(1)}$ is given, for example, by 3σ (Eq. 4.13). Thus, the error in the site occupation fraction is evaluated by

$$\Delta k = \left(\sum_{i,n} \left(\Delta k_i^{(n)}\right)^2\right)^{\frac{1}{2}} \quad (4.27)$$

Table 4.5. Atomic ratios of constituent elements at the α and β planes for $Ti_{43}Al_{55}Nb_2$.

	α Plane	β Plane
Ti	x	0
Al	$y - w$	$w = 0.5 - z \cdot (1 - k)$
Zr	$z \cdot k$	$z \cdot (1 - k)$

$x = 0.43$; $y = 0.55$; $z = 0.02$

Table 4.6. X-ray intensities used for ALCHEMI on $Ti_{43}Al_{55}Nb_2$.

	Diffraction condition		
	$n = 1$ (s > 0)	$n = 2$ (s < 0)	$n = 3$ (nonchanneling)
$N_{Ti}^{(n)}$	56778	100267	64867
$N_{Al}^{(n)}$	51368	55543	48485
$N_{Nb}^{(n)}$	4067	7042	4370

In addition to TiAl with the $L1_0$-type structure noted above, the application to various structures such as Ni_3Al [30,31] with $L1_2$-type structure, TiNi [32] with B2-type structure, and so on [33,34] have been widely carried out.

If the structure factor can be evaluated from other methods such as the intersecting Kikuchi-line (IKL) method [35], the site occupancy of the constituent elements can be determined without assuming that the host elements form the specific sublattice [36].

4.6.3 Accuracy and Notice of ALCHEMI

4.6.3.1 Accuracy of ALCHEMI

Crystal Structure. The accuracy of ALCHEMI strongly depends on the crystalline structure of a material. In the case of Ni_3Al with the $L1_2$-type structure shown in Fig. 4.19b, there are two atomic planes parallel to the electron beam. One consists of Ni atoms and the other of Ni and Al atoms. Eventually, the intensity of the superlattice reflection is weak, and the electron channeling effect is also weak. Thus, the accuracy of ALCHEMI for a structure such as the $L1_2$-type structure of Ni_3Al is low.

ALCHEMI with an axial incidence such as along the [001] direction was carried out in Ni_3Al [37]. With such an axial channeling condition, attention should be paid to the delocalization effect noted below.

Crystallinity. When a specimen has a lot of strain, the diffraction condition changes from place to place. Thus, the channeling effect cannot be strongly induced, and the accuracy of the analysis generally is low. ALCHEMI should be carried out in such a region where the Kikuchi lines are clearly observed, as shown in Fig. 4.20. To reduce thermal diffuse scattering, ALCHEMI has been carried out at low temperatures [33].

Host Elements. The electron channeling effect is strongly induced when the intensity of the superlattice reflection is strong. Thus, when the difference in atomic numbers of host elements is large, the accuracy of ALCHEMI is high.

Impurities (Additives). The statistical errors shown in Eq. 4.27 are mainly introduced by characteristic X-rays from a small amount of impurity. Because the fluorescent yield increases with an increase in atomic number, as noted in Section 4.1, the site location of the impurity of a large atomic number can generally be determined with high accuracy.

4.6.3.2 Notice of ALCHEMI

Delocalization Effect. For derivation of the site occupation shown in Section 4.6.2, it is assumed that changes in the characteristic X-ray intensity depend on the distribution of the incident electron density alone. In other words, spreading of the atomic electrons that generate the characteristic X-ray is neglected, and the atomic electrons are assumed to be localized as a δ-function. However, in the axial illumination condition, where many reflections are excited simultaneously, the density of incident electrons has sharp peaks; thus, the difference in the spreading of atomic electrons for constituent elements (the so-called delocalization effect) should be taken into account when determining site occupancy [38,39].

ALCHEMI with EELS. ALCHEMI utilizes a strong interaction between the incident electrons and materials and analyzes the resultant characteristic X-rays to determine site occupancy. During the analysis it is assumed that the characteristic X-rays emitted do not strongly interact with matter. The principle of ALCHEMI can be utilized with the core loss in electron energy-loss spectra. However, in this case, attention should be paid to the interaction between the inelastically scattered electrons and the material. If there is a diffraction effect of the inelastic electrons in the crystal, quantitative analysis of the site occupation is not easy.

Appendix

The derivation of Eq. 4.22 for site occupancy under three diffraction conditions is described in the following.

First, eliminate Pc from Eq. 4.19

$$P_C = N_C^{(n)} \big/ \left(C_\alpha \cdot I_\alpha^{(n)} + C_\beta \cdot I_\beta^{(n)}\right) \tag{4.28}$$

by inserting the diffraction conditions $n = 1, 2$

$$N_C^{(1)} \big/ \left(C_\alpha \cdot I_\alpha^{(1)} + C_\beta \cdot C_\beta^{(1)}\right) = N_C^{(2)} \big/ \left(C_\alpha \cdot I_\alpha^{(2)} + C_\beta \cdot I_\beta^{(2)}\right) \tag{4.29}$$

$$N_C^{(1)} \big/ \left(C_\alpha \cdot I_\alpha^{(2)} + C_\beta \cdot I_\beta^{(2)}\right) = N_C^{(2)} \big/ \left(C_\alpha \cdot I_\alpha^{(1)} + C_\beta \cdot I_\beta^{(1)}\right) \tag{4.30}$$

and modifying the equation

$$\frac{N_C^{(1)}}{N_C^{(2)}}\left(C_\alpha + C_\beta \cdot \frac{I_\beta^{(2)}}{I_\alpha^{(2)}}\right) = \frac{I_\alpha^{(1)}}{I_\alpha^{(2)}}\left(C_\alpha + C_\beta \cdot \frac{I_\beta^{(1)}}{I_\alpha^{(1)}}\right) \tag{4.31}$$

Here, with $N_C = N_C^{(1)}/N_C^{(2)}$, $R_{\beta\alpha}^{(n)} = I_\beta^{(n)}/I_\alpha^{(n)}$, $R_\alpha = I_\alpha^{(1)}/I_\alpha^{(2)}$, the equation is simplified as

$$N_C\left(C_\alpha + C_\beta \cdot R_{\beta\alpha}^{(2)}\right) = R_\alpha\left(C_\alpha + C_\beta \cdot R_{\beta\alpha}^{(1)}\right) \tag{4.32}$$

When the fraction of the C atom in the α-plane is k, C_α and C_β are given as $C_\alpha = z \cdot k$ and $C_\beta = z(1 - k)$, respectively. By inserting these terms in Eq. 4.32

$$N_C\left(z \cdot k + z(1-k) \cdot R_{\beta\alpha}^{(2)}\right) = R_\alpha\left(z \cdot k + z(1-k) \cdot R_{\beta\alpha}^{(1)}\right) \tag{4.33}$$

Equation 4.22 for site occupancy k is obtained as

$$k = \frac{R_\alpha \cdot R_{\beta\alpha}^{(1)} - N_C \cdot R_{\beta\alpha}^{(2)}}{N_C\left(1 - R_{\beta\alpha}^{(2)}\right) - R_\alpha\left(1 - R_{\beta\alpha}^{(1)}\right)}$$

Now, $R_{\beta\alpha}^{(1)}$, $R_{\beta\alpha}^{(2)}$, and R_α is obtained. To do that, first $P_A/P_B = P_{AB}$ is obtained. Here, we use data from the nonchanneling condition. From (Eq. 4.17) / (Eq. 4.18) with $n = 3$

$$\frac{N_A^{(3)}}{N_B^{(3)}} = \frac{P_A}{P_B} \cdot \frac{A_\alpha \cdot I_\alpha^{(3)} + A_\beta \cdot I_\beta^{(3)}}{B_\alpha \cdot I_\alpha^{(3)} + B_\beta \cdot I_\beta^{(3)}} = \frac{P_A}{P_B} \cdot \frac{A_\alpha + A_\beta}{B_\alpha + B_\beta} \tag{4.34}$$

because for the nonchanneling condition, the densities of incident electrons for the α- and β-planes are equal and so $I_\alpha^{(3)} = I_\beta^{(3)}$. Then P_{AB} can be given as

$$P_{AB} = \frac{P_A}{P_B} = \frac{N_A^{(3)}}{N_B^{(3)}} \cdot \frac{B_\alpha + B_\beta}{A_\alpha + A_\beta} \tag{4.35}$$

Now, with the data of the channeling condition (n), $R_{\beta\alpha}^{(n)}$ $(= I_\beta^{(n)}/R_\alpha^{(n)})$ is obtained. From (Eq. 4.17) / (Eq. 4.18)

$$\frac{N_A^{(n)}}{N_B^{(n)}} = \frac{P_A}{P_B} \cdot \frac{A_\alpha \cdot I_\alpha^{(n)} + A_\beta \cdot I_\beta^{(n)}}{B_\alpha \cdot I_\alpha^{(n)} + B_\beta \cdot I_\beta^{(n)}} = \frac{P_A}{P_B} \cdot \frac{A_\alpha + A_\beta \cdot I_\beta^{(n)}/I_\alpha^{(n)}}{B_\alpha + B_\beta \cdot I_\beta^{(n)}/I_\alpha^{(n)}} = \frac{P_A}{P_B} \cdot \frac{A_\alpha + A_\beta \cdot R_{\beta\alpha}^{(n)}}{B_\alpha + B_\beta \cdot R_{\beta\alpha}^{(n)}} \tag{4.36}$$

By inserting Eq. 4.35

$$\frac{N_A^{(n)}}{N_B^{(n)}}\left(B_\alpha + B_\beta \cdot R_{\beta\alpha}^{(n)}\right) = P_{AB}\left(A_\alpha + A_\beta \cdot R_{\beta\alpha}^{(n)}\right) \tag{4.37}$$

$$R_{\beta\alpha}^{(n)}\left(B_\beta \frac{N_A^{(n)}}{N_B^{(n)}} - A_\beta \cdot P_{AB}\right) = A_\alpha \cdot P_{AB} - B_\alpha \cdot \frac{N_A^{(n)}}{N_B^{(n)}} \tag{4.38}$$

$$R_{\beta\alpha}^{(n)} = \frac{A_\alpha \cdot P_{AB} - B_\alpha \cdot \dfrac{N_A^{(n)}}{N_B^{(n)}}}{B_\beta \cdot \dfrac{N_A^{(n)}}{N_B^{(n)}} - A_\beta \cdot P_{AB}} \tag{4.39}$$

Finally, $R_\alpha(= I_\alpha^{(1)}/R_\alpha^{(2)})$ is obtained. From Eq. 4.17

$$\begin{aligned}\frac{N_A^{(1)}}{N_A^{(2)}} &= \frac{P^A\left(A_\alpha \cdot I_\alpha^{(1)} + A_\beta \cdot I_\beta^{(1)}\right)}{P_A\left(A_\alpha \cdot I_\alpha^{(2)} + A_\beta \cdot I_\beta^{(2)}\right)} \\ &= \frac{I_\alpha^{(1)}}{I_\alpha^{(2)}} \cdot \frac{A_\alpha + A_\beta \cdot I_\beta^{(1)} / I_\alpha^{(1)}}{A_\alpha + A_\beta \cdot I_\beta^{(2)} / I_\alpha^{(2)}} \\ &= R_\alpha \cdot \frac{A_\alpha + A_\beta \cdot R_{\beta\alpha}^{(1)}}{A_\alpha + A_\beta \cdot R_{\beta\alpha}^{(2)}}\end{aligned} \tag{4.40}$$

$$R_\alpha = \frac{N_A^{(1)}}{N_A^{(2)}} \cdot \frac{A_\alpha + A_\beta \cdot R_{\beta\alpha}^{(2)}}{A_\alpha + A_\beta \cdot R_{\beta\alpha}^{(1)}} \tag{4.41}$$

References

1. Nomenclature, symbols, units and their usage in spectrochemical analysis-VIII. Nomenclature system for X-ray spectroscopy. Recommendations (1991)
2. Jenkins R, Manne R, Robin J, Senemaud C (1991) Part VIII. Nomenclature system for X-ray spectroscopy. Pure Appl Chem 63:735
3. Wollman DA, Irwin KD, Hilton GC, Dulcie LL, Newbury DE, Martinis JM (1997) High-resolution, energy-dispersive microcalorimeter spectrometer for X-ray microanalysis. J Microsc 188:196
4. Wollman DA, Irwin KD, Hilton GC, Dulcie LL, Bergren NF, Newbury DE, Martinis JM (1998) Microcalorimeter EDS with 3 eV energy resolution. In: Proceedings of the 14th international conference on electron microscopy, vol 3, p 573
5. Zaluzec NJ (1979) Quantitative X-ray microanalysis. In: Introduction to analytical electron microscopy. Hen JJ, Goldstein JI, Joy DC. (Plenum, New York, p 121)
6. Yang J-M, Shindo D, Takeguchi M, Kawasaki M, Oikawa T (1999) Characterization of microstructure and magnetic domain structure in Sm-Co based permanent magnets by advanced transmission electron microscopy. J Jpn Inst Metals 63:542 (In Japanese)
7. Ziebold TO (1967) Precision and sensitivity in microprobe analysis. Anal Chem 39:858
8. Watanabe M, Williams DB (1999) Atomic-level detection by X-ray microanalysis in the analytical electron microscope. Ultramicroscopy 78:89
9. Kawasaki M, Oikawa T, Ibe K, Park K-H, Shiojiri M (1998) EDS elemental mapping of a DRAM with an FE-TEM. J Electron Microsc 47:335
10. Cliff G, Lorimer GW (1975) The quantitative analysis of thin specimens. J Microsc 103:203
11. Schreiber TP, Wims AM (1981) A quantitative X-ray microanalysis thin film method using K-, L-, and M-lines. Ultramicroscopy 6:323
12. Goldstein JI, Williams DB, Cliff G (1986) Quantitative X-ray analysis. In: Joy DC, Romig AD Jr, Goldstein JI (eds) Principles of analytical electron microscopy. (Plenum, New York, p 155)
13. Mott NF, Massey HSW (1949) The theory of atomic collisions, 2nd edn. Oxford University Press, London, p 243
14. Green M, Cosslett VE (1961) The efficiency of production of characteristic X-radiation in thick targets of a pure element. Proc Phys Soc 78:1206
15. Horita Z (1998) Quantitative X-ray microanalysis in analytical electron microscopy. Mater Trans JIM 39:947
16. Überall H (1956) High-energy interference effect of bremsstrahlung and pair production in crystals. Phys Rev 103:1055
17. Barbiellini G, Bologna G, Diambrini G, Murtas GP (1962) Experimental evidence for a quasi-monochromatic bremsstrahlung intensity from the Frascati 1-GeV electronsynchrotron. Phys Rev Lett 8:454
18. Spence JCH, Reese G, Yamamoto N, Kurizki G (1983) Coherent bremsstrahlung peaks in X-ray microanalysis spectra. Phil Mag B48:L39
19. Reese GM, Spence JCH, Yamamoto N (1984) Coherent bremsstrahlung from kilovolt electrons in zone axis orientations. Phil Mag A49:697
20. Spence JCH, Reese G (1986) Pendellösung radiation and coherent bremsstrahlung. Acta Cryst A42: 577
21. Satoh T, Otsuki E, Shindo D (1998) Coherent bremsstrahlung in ferrite observed by an analytical transmission electron microscope. J Electron Microsc 47:345
22. Shindo D, Hiraga K, Williams T, Hirabayashi M, Inoue A, Masumoto T (1989) Electron channelling effect in an Al-Fe-Cu quasicrystal. Jpn J Appl Phys 28:L688
23. Cowley JM (1964) The derivation of structural information from absorption effects in X-ray diffraction. Acta Cryst 17:33
24. Batterman BW (1969) Detection of foreign atom sites by their X-ray fluorescence scattering. Phys Rev Lett 22:703
25. Spence JCH, Taftø J (1983) ALCHEMI: a new technique for locating atoms in small crystals. J Microsc 130:147

26. Shindo D, Hirabayashi M, Kawabata T, Kikuchi M (1986) A channelling enhanced microanalysis on niobium atom location in an Al-43%Ti-2%Nb intermetallic compound. J Electron Microsc 35:409
27. Shindo D, Chiba A, Hiraga K, Hanada S (1991) Electron channelling enhanced microanalysis of intermetallic compounds. In: Izumi O (ed) Proceedings of the International Symposium on Intermetallic Compounds, p 87
28. Horita Z, Matsumura S, Baba T (1995) General formulation for ALCHEMI. Ultramicroscopy 58:327
29. Rossouw CJ, Forwood CT, Gibson MA, Miller PR (1996) Statistical ALCHEMI: general formulation and method with application to Ti-Al ternary alloys. Phil Mag A74:57
30. Shindo D, Kikuchi M, Hirabayashi M, Hanada S, Izumi O (1988) Site determination of Fe, Co and Cr atoms added in Ni_3Al by electron channelling enhanced microanalysis. Trans Jpn Inst Metall 29:956
31. Chiba A, Shindo D, Hanada S (1991) Site occupation determination of Pd in Ni_3Al by ALCHEMI. Acta Metall Mater 39:13
32. Nakata Y, Tadaki T, Shimizu K (1991) Atom location of the third element in Ti-Ni-X shape memory alloys determined by the electron channelling enhanced microanalysis. Mater Trans JIM 32:580
33. Spence JCH, Graham RJ, Shindo D (1986) Cold ALCHEMI: impurity atom site location and the temperature dependance of dechannelling. Mater Res Soc Symp Proc 62:153
34. Okaniwa H, Shindo D, Yoshida M, Takasugi T (1999) Determination of site occupancy of additives X (X = V, Mo, W and Ti) in the Nb-Cr-X Laves phase by ALCHEMI. Acta Mater 47:1987
35. Gjønnes J, Høier R (1971) The application of non-systematic many-beam dynamical effects to structure factor determination. Acta Cryst A27:313
36. Matsumura S, Morimura T, Oki K (1991) An analytical electron diffraction technique for the determination of long-range order parameters in multi-component ordered alloys. Mater Trans JIM 32:905
37. Bentley J (1986) Axial electron channeling microanalysis of Ll_2 ordered alloys. In: Proceedings of the 11th International Congress on Electron Microscopy, Kyoto, vol 1, p 551
38. Pennycook SJ (1985) Electron channeling analysis and Z-contrast imaging of dopants in semiconductors. In: Bailey GW (ed) Proceedings of the 43rd annual EMSA meeting. San Francisco Press, San Francisco, p 296
39. Pennycook SJ (1988) Delocalization corrections for electron channeling analysis. Ultramicroscopy 26:239

5. Peripheral Instruments and Techniques for Analytical Electron Microscopy

5.1 Electron Diffraction

Electron diffraction is fundamentally one of the most important methods for obtaining crystallographic information about materials. The information obtained by electron diffraction is the quantity in reciprocal space, which is the same as that attained by X-ray diffraction and neutron diffraction. Although the intensity of X-ray diffraction and neutron diffraction directly corresponds to the square of an absolute value of the structure factor according to the *kinematical diffraction theory*, the intensity of electron diffraction should be interpreted on the basis of the *dynamical diffraction theory*. The dynamical diffraction effect on electron diffraction is explained in the literature [1–3]. In this chapter we discuss the principles and application of nano-beam electron diffraction and convergent beam electron diffraction, which extensively utilize the function of an analytical electron microscope.

5.1.1 Nano-beam Electron Diffraction

5.1.1.1 Limitation of Selected Area Electron Diffraction

The formation process of an electron diffraction pattern is described in Section 2.1.5. The method by which an electron diffraction pattern is obtained with an aperture in the objective image plane illuminating a wide area (see Fig. 2.10a) is called *selected area electron diffraction*. The aperture selecting the specific area is called a *selected area aperture*. With this method the diffraction pattern is obtained from an area selected by the selected area aperture. The smallest diameter of the aperture is about 5 μm owing to the accuracy of mechanical fabrication. The magnification of the objective lens is generally about 50×; therefore, the smallest diameter d_a of the selected area in the specimen plane is estimated to be

$$d_a = 5\,\mu\text{m}/50 = 0.1\,\mu\text{m} \qquad (5.1)$$

This is the smallest area that can be analyzed by selected area electron diffraction.

5.1.1.2 Nano-beam Electron Diffraction

When a small probe is formed with the condenser lens system (see Fig. 2.10c, NBD mode), maintaining the diffraction mode in the image-forming lens system, an electron diffraction pattern can be obtained from the area illuminated with the probe (Fig. 5.1a). Because the probe size can be reduced to nanometer order, the method is called *nano-beam electron diffraction*. The selected area aperture is not used with this method. The method is effective for structural analysis of precipitates and interfaces in a small area.

5.1.1.3 Notes on Nano-beam Electron Diffraction

The ideally small and parallel electron illumination is illustrated in Fig. 5.1a. In practice, however, the focused electrons are illuminated on a finite area of a specimen. In this case, the spots of the diffraction pattern become disks corresponding to the illumination angle (α). Therefore, the illumination angle should be minimized by selecting a condenser aperture with a smaller diameter to obtain disks of smaller diameter. The diffraction focus (focus of the first intermediate lens) should be adjusted at the back-focal plane of the objective lens to obtain the exact camera length. To adjust the diffraction focus, it is convenient to adjust it in the selected area diffraction mode with the parallel illumination condition in advance.

5.1.2 Convergent Beam Electron Diffraction

5.1.2.1 Principle

Whereas diffraction spots are observed with parallel illumination, disks are observed in the diffraction pattern when the electron beam converges on the specimen, as shown in Fig. 5.1b. This is the *convergent beam electron diffraction*

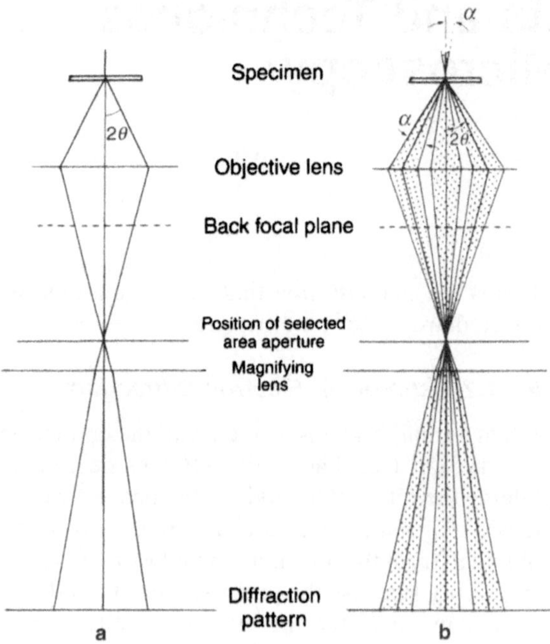

Fig. 5.1. Ray diagrams of nano-beam diffraction **a** and convergent beam electron diffraction (CBED) **b**

(CBED) pattern. CBED is different from nano-beam diffraction, and the information of scattered electrons is obtained as a function of the incidence angle against the specimen. In other words, the intensity distribution of the diffracted beam for the change in the deviation parameter, which indicates the excitation error, can be obtained. The convergence angle α is given as

$$\alpha = \frac{\lambda r}{dR} \tag{5.2}$$

where λ is the wavelength of the electron, and d is the spacing of the lattice plane. The parameters r and R are measured on a film or a print; r corresponds to the radius of the diffraction disk, and R is the distance between the transmitted beam and the reciprocal lattice point, corresponding to $1/d$. Crystal symmetry, polarity, and crystal thickness can be obtained from a CBED pattern, as explained below.

5.1.2.2 Application of CBED

Determination of Crystal Symmetry. When the crystal symmetry is determined from CBED patterns, a parallel-sided area in a specimen should be found. This is because the intensity profile in a CBED pattern contains information about crystal thickness, and so precise determination of crystal symmetry is difficult for such regions with inhomogeneous thickness. As shown in Table 5.1, there are 10 symmetry elements that keep the parallel-sided crystal unchanged: 6 two-dimensional symmetry elements and 4 three-dimensional elements.

Table 5.1. Symmetry elements in a parallel-sided plate.

Two-dimensional symmetry element
One-fold rotation axis
Two-fold rotation axis
Three-fold rotation axis
Four-fold rotation axis
Six-fold rotation axis
Vertical mirror plane
Three-dimensional symmetry element
Horizontal mirror plane
Inversion center
Horizontal two-fold rotation axis
Four-fold rotatory inversion

Figure 5.2a shows a CBED pattern of Si with the [111] incidence. Its contour map in Fig. 5.2b shows the intensity distribution in detail. The intensity distribution around the transmitted beam displays six-fold symmetry, resulting from information of the zero-order Laue zone or two-dimensional information projected along the incident electron beam. However, the detailed intensity distribution in the disk of the transmitted beam (e.g., the Kikuchi lines) corresponding to the higher-order Laue zone, as indicated by arrows, shows three-fold symmetry. The latter finding results from the three-dimensional symmetry element, and eventually the CBED pattern of Fig. 5.2b indicates the existence of the three-fold symmetry axis and the mirror plane.

In addition to a CBED pattern with the incidence along the zone axis, as shown in Fig. 5.2a, other CBED patterns should be observed under the two-beam conditions ($+g$ excitation and $-g$ excitation) to determine the diffraction group. A total of 31 diffraction groups and the symmetries in the corresponding CBED patterns are given by Buxton et al. [4].

In general, the space groups can be determined as follows: First, CBED patterns are observed with plural crystal orientations. The diffraction group can be obtained based on the symmetry of each CBED pattern. Eventually, the true point group for the 32 kinds that satisfies all the diffraction groups observed can be determined. The relation between the diffraction group and the point group can be also found in Buxton et al.'s paper [4]. Finally, a space group among the 230 kinds can be

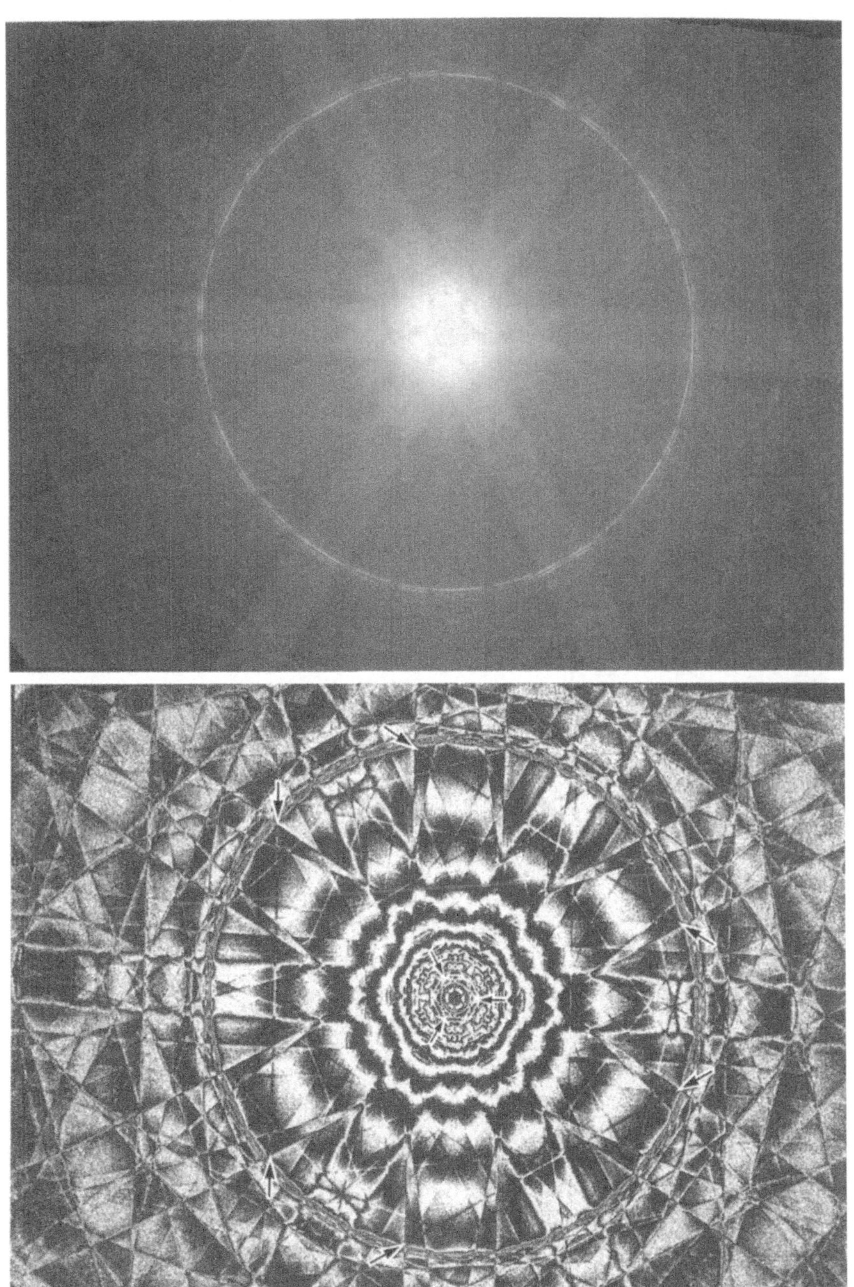

Fig. 5.2. CBED pattern of Si with [111] incidence **a** and its contour map **b**

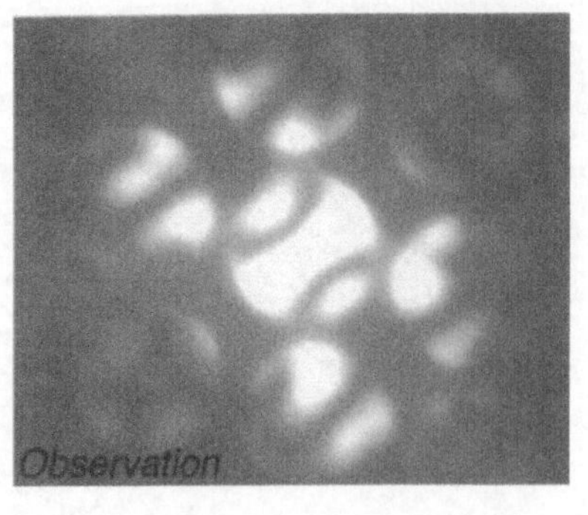

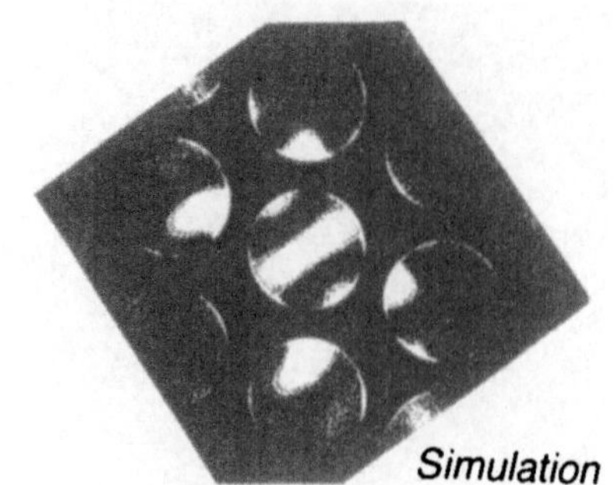

Fig. 5.3. CBED pattern of GaAs with [01$\bar{1}$] incidence and its simulation (*top right*). Structure model and crystal orientations are at the bottom. Observed and calculated CBED patterns are kindly provided by Drs. I. Yonenaga and W.G. Burgess [6]

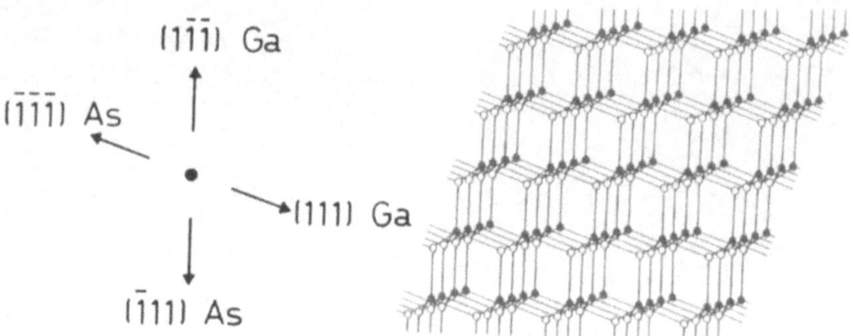

determined from the point group and the extinction rule appearing in electron diffraction patterns [5].

In relation to the symmetry dermination, the polarity of a crystal without a center of symmetry can be determined with CBED. Figure 5.3a,b shows a CBED pattern of GaAs with [01$\bar{1}$] incidence and its simulation. The crystal orientations and the atomic arrangement are shown in Fig. 5.3c,d. From the intensity distribution in the disks, it is seen that the crystal has a mirror symmetry along the axis parallel to the [100] direction, but it does not have the symmetry along the axis parallel to the [0$\bar{1}\bar{1}$] direction. The polarity of this crystal can be determined in this way. Determining the polarity is important for understanding the crystal growth process and determining the dislocation type [6].

Evaluation of Specimen Thickness. As noted in Section 3.5.2, specimen thickness can be evaluated by electron energy-loss spectroscopy (EELS), but it is necessary to know the mean free path for inelastic scattering. On the other hand, with CBED, the thickness can be evaluated accurately with a single CBED pattern [7, 8].

Principle. Under the two-beam condition, the diffraction intensity of the reflection g for thickness t is given as

$$|\Phi_g|^2 = \sin^2\beta \cdot \sin^2 \pi\Delta k t \quad (5.3)$$

where

$$\beta = \tan^{-1}\left(\frac{1}{s\xi_g}\right) \quad (5.4)$$

$$\Delta k = \frac{\sqrt{1+(s\xi_g)^2}}{\xi_g} \quad (5.5)$$

where s is the deviation parameter, and ξ_g is the extinction distance for the reflection g. As shown in Fig. 5.4 and using Eq. 5.2, s is given with α_s as

$$s = g\alpha_s = \frac{\lambda r_s}{d^2 R} \quad (5.6)$$

where r_s is a distance measured in the diffraction disk. The conditions of Eq. 5.3 for taking the minimum and maximum values can be obtained by differentiating Eq. 5.3 as

$$\Delta k t = \text{integer} \quad (5.7)$$

$$\tan \pi\Delta k t = \pi\Delta k t \quad (5.8)$$

When $\Delta k \cdot t$ satisfying Eqs. 5.7 and 5.8 is set to be n_k and x_k respectively, n_k and x_k for a positive integer k are given in Table 5.2. Now k can be given with $k = i + j$. Here, i indicates ith intensity minima and maxima from the center ($s = 0$) of the diffraction disk g, and j is an integer (≥ 0) that

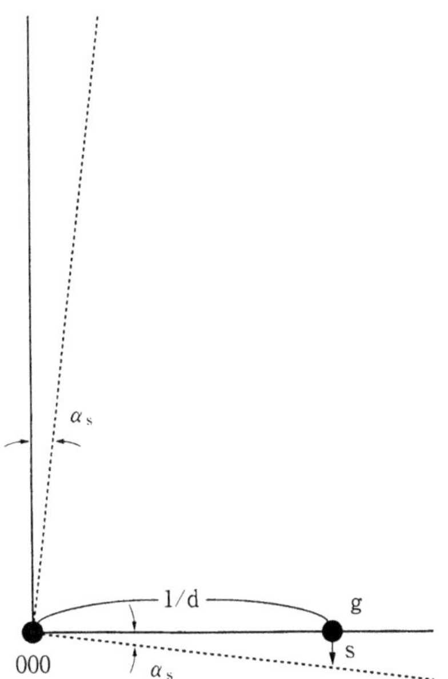

Fig. 5.4. Relation of a deviation parameter (s) and convergence angle α_s

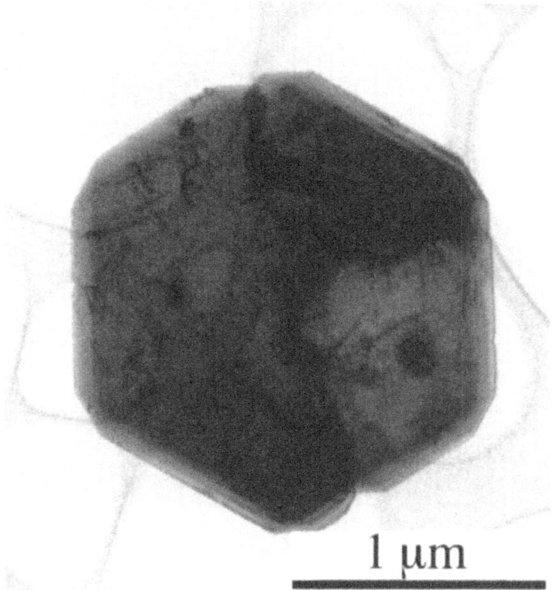

Fig. 5.5. Transmission electron microscope (TEM) image of a platelet-type hematite (α-Fe_2O_3) particle

Table 5.2. Values of n_k and x_k for a positive integer k.

	Values for k				
Parameter	1	2	3	4	5
n_k	1	2	3	4	5
x_k	1.431	2.459	3.471	4.477	5.482

depends on a value of t/ξ_g [8]. By using Eq. 5.5, the relation between s_i and n_k or $x_{k'}$ is given as

$$(s_i/n_k)^2 = -(1/\xi_g)^2(1/n_k)^2 + (1/t)^2 \qquad (5.9)$$

$$(s_i/x_{k'})^2 = -(1/\xi_g)^2(1/x_{k'})^2 + (1/t)^2 \qquad (5.10)$$

Thus, in Eq. 5.9, for example, if an appropriate integer j is selected for the specimen thickness, the linear relation between $(s_i/n_k)^2$ and $(1/n_k)^2$ can be obtained by evaluating s_i, which is the deviation parameter in the ith intensity minima. The slope of the line is $-(1/\xi_g)^2$, and the intercept of the line in the axis of $(s_i/n_k)^2$ is $(1/t)^2$. The thickness is evaluated from the intercept. For the intensity maxima of Eq. 5.10, the linear relation between $(s_i/n_k')^2$ and $(1/x_k')^2$ can be obtained for the thickness determination in the same way.

Practice. An example of thickness evaluation is given below with a platelet-type hematite (α-Fe_2O_3) particle shown in Fig. 5.5. The hematite has a corundum structure, and its CBED pattern is shown in Fig. 5.6. The CBED pattern was observed under the two-beam condition with the $03\bar{3}0$ reflection excited. By evaluating the distances r_s ($= \Delta\theta_i$) and R ($= 2\ \theta_g$), s_i can be evaluated from Eq. 5.6. In principle, j cannot be determined without knowledge of the specimen thickness. Now, if the relation between $(s_i/n_k)^2$ and $(1/n_k)^2$ is plotted inserting various integers into j, the linear relation is found only when a correct interger is assigned to j. Thus, the thickness is evaluated from the intercept of the line selected. Figure 5.7 shows the relation between $(s_i/n_k)^2$ and $(1/n_k)^2$ for the data of Fig. 5.6 assuming $j = 0, 1, 2, 3$. The straight line is obtained for $j = 2$, and the thickness is evaluated to be $t = 171$ nm from the intercept [9]. The intensity minima were analyzed in Fig. 5.6, but the intensity maxima can be also utilized to determine the thickness with higher accuracy.

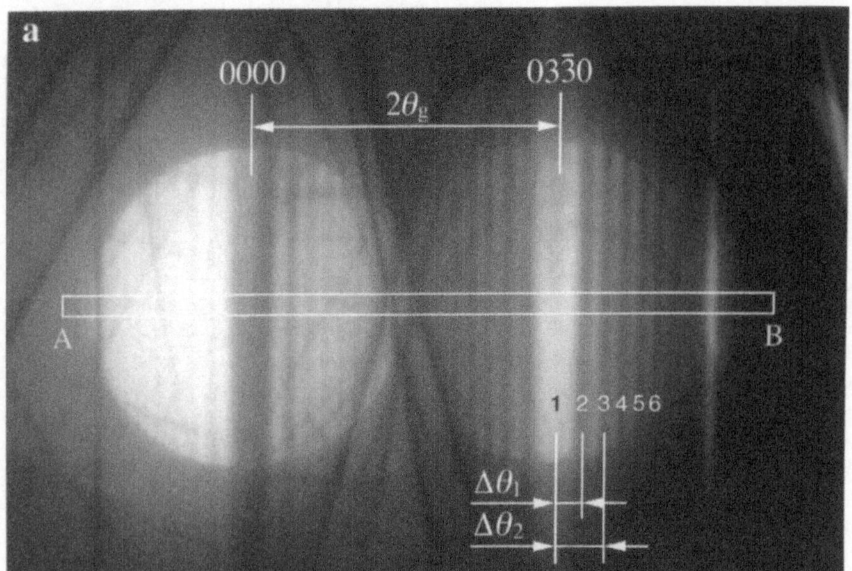

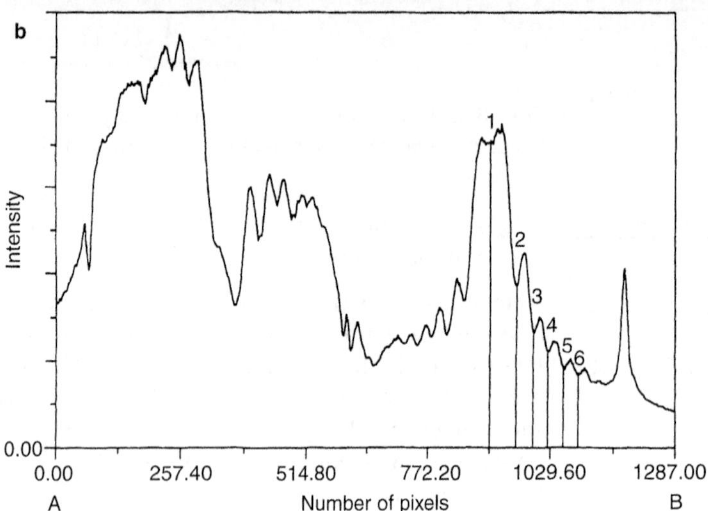

Fig. 5.6. CBED pattern of a platelet-type hematite particle under the two-beam condition **a** and its intensity profile **b**

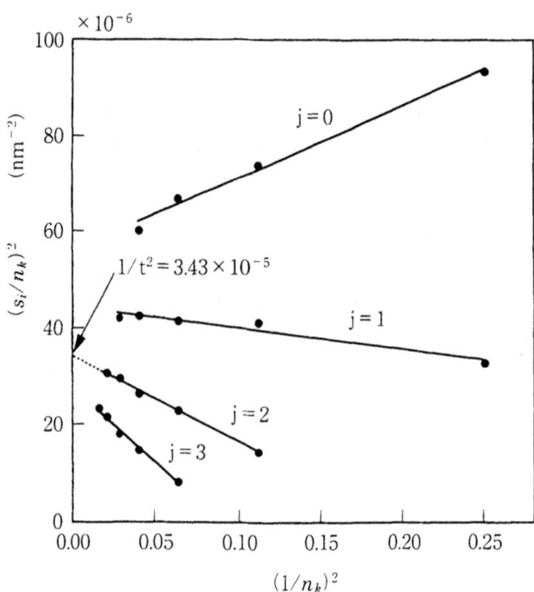

Fig. 5.7. Relation between $(s_i/n_k)^2$ and $(1/n_k)^2$ with the data of Fig. 5.6 for $j = 0, 1, 2, 3$

Bragg Condition and Higher-order Diffraction

In general, diffraction occurs when the following Bragg condition is satisfied:

$$2d\sin\theta = n\lambda \quad (5.11)$$

where d and θ are the spacing of the lattice plane and the diffraction angle (2θ is the scattering angle); λ is a wavelength of the electron; and n is a positive integer showing the nth order diffraction. Thus, Eq. 5.11 indicates that the nth order diffraction occurs when the electron of wavelength λ is incident on a crystal of a spacing in the lattice plane d. On the other hand, Eq. 5.11 can be modified as

$$2\left(\frac{d}{n}\right)\sin\theta = \lambda \quad (5.12)$$

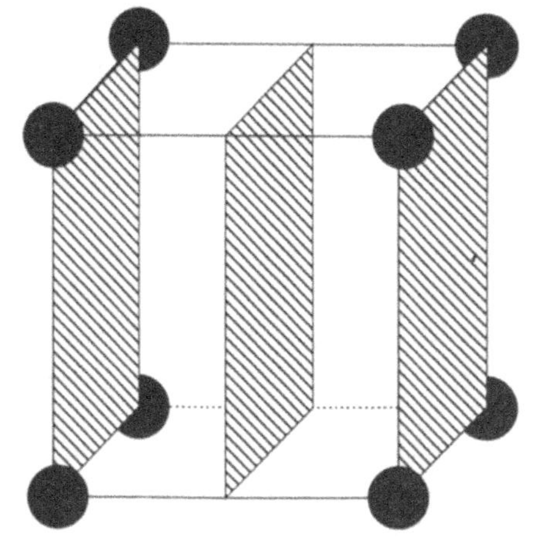

Fig. 5.8. Artificial lattice planes with the spacing of d/n ($n = 2$)

Equation 5.12 indicates that the nth order diffraction noted above can be interpreted as the first-order diffraction from the artificial lattice plane whose spacing is d/n, as shown in Fig. 5.8. Thus, if the spacing of lattice plane d/n can be redefined as d, the Bragg condition can be written as

$$2d\sin\theta = \lambda$$

This equation is frequently used as a conventional equation, as shown in Eq. 2.5.

5.1.2.3 Evaluation of Lattice Spacing and Lattice Strain with HOLZ Patterns

In the CBED pattern of Fig. 5.2a, the bright region of the transmitted beam and the $2\bar{2}0$ reflection corresponds to the zero-order Laue zone (ZOLZ). A white ring consisting of sharp bright Kikuchi lines is observed in the outer region. This region corresponds to the first-order Laue zone (FOLZ), where the diffraction disks are not observed and only the Kikuchi lines in the disks are sharply visible, as the diffraction intensity drastically decreases with the increase of excitation error. It is known that Kikuchi lines generally appear as a pair of black and white parallel lines; thus in Fig. 5.2a bright Kikuchi lines appear in the FOLZ, and black ones appear in the ZOLZ. In general, for lower accelerating voltage the second-order Laue zone (SOLZ) appears outside the FOLZ, and a third-order Laue zone (TOLZ) may be seen in an outer region. In the transmitted beam and its peripheral region, many dark Kikuchi lines appear in correspondence with the bright rings in the higher-order Laue zone. Kikuchi lines and patterns resulting from the effect of the higher-order Laue zone are called *HOLZ (higher-order Laue zone) lines* and *patterns*, respectively. The pattern resulting from the effect of the zero-order Laue zone is called a *ZOLZ pattern*. ZOLZ patterns reflect two-dimensional projected crystallographic information, and HOLZ patterns provide three-dimensional crystallographic information.

A HOLZ pattern is sensitive to the change of a lattice constant, and thus the lattice constant can be determined with a HOLZ pattern when the accelerating voltage is known. In practice, a HOLZ pattern can be simulated on the basis of kinematical approximation, and the lattice constant can be determined by comparing the simulated pattern with the observed pattern [5]. The small change of a lattice constant due to a temperature change or the existence of a lattice defect can also be detected with HOLZ patterns. The Burgers vector of a dislocation can be determined with the CBED method [10]. Furthermore, the structure factor can also be determined by measuring the separation of Kikuchi lines [11] or the intensity change of the Kikuchi lines for the variation in accelerating voltage [12].

5.2 Lorentz Microscopy

Because the principles of imaging the magnetic domain structure can be understood in terms of the Lorentz force on electrons, electron microscopy for observing magnetic domains by utilizing the *Lorentz force* is known as *Lorentz microscopy* [1, 13]. Lorentz microscopy with a conventional TEM consists of the *Fresnel method* and the *Foucault method*, which are utilized under defocused and just-focused conditions, respectively.

In general, the magnetic field at the specimen position in a TEM is about 1200 kA/m, corresponding to a magnetic flux density of 1.5 T; this strong magnetic field destroys or modifies the inherent magnetic domain structure. Thus, the magnetic field at the specimen position should be reduced especially for observing soft magnetic materials. One of the easiest ways to reduce the magnetic field is to switch off and degauss the objective lens. Usually with these adjustments the residual magnetic field can be reduced to less than 160 A/m (0.2 mT). However, fine TEM images are not expected under this condition, as the objective lens, which determines the resolution of a TEM, is not used. For keeping the action of the objective lens to observe detailed magnetic structure, the position of the specimen should be shifted, leaving the strong magnetic field; or a special shield for the magnetic field should be introduced in the objective lens.

Specimens should be thin for observing TEM images; thus the magnetic domain structure is considered to be basically different from that in bulk materials. It is thought that in thin films Néel walls are observed, whereas in relatively thick films Bloch walls tend to appear [14]; between these thickness ranges cross-tie walls appear [15, 16]. Thus, the specimen should be as thick as possible. From this point of view, a high-voltage TEM is useful, as relatively thick specimens can be observed with high-energy electrons. For studying magnetic domain structure with Lorentz microscopy, as described here, a high-voltage electron microscope (JEM-ARM 1250) with an accelerating voltage of 1250 kV is utilized. Because the position of the specimen in the high-voltage electron microscope can be shifted to a higher position above the objective lens, the magnetic field at this position is reduced to about 400 A/m (0.5 mT).

5.2.1 Principles

The Lorentz force for an electron passing through a specimen with the magnetic flux density $\vec{B}$ given by

$$\vec{F} = -e(\vec{v} \times \vec{B}) \tag{5.13}$$

where v is the electron velocity, and e is the elementary electrical charge. The direction of the force expressed by the vector product in Eq. 5.13 is given by Fleming's left-hand rule. Thus, the Lorentz force on an electron passing through a magnetized specimen, which has the magnitude of magnetization $\vec{I}_s$, is given by

$$\vec{F} = -e(\vec{v} \times \vec{I}_s) \tag{5.14}$$

Here, the direction of the magnetization in the specimen of thickness t is assumed to be normal to the plane of the diagram (y direction) as shown in Fig. 5.9. By magnetization $\vec{I}_s$, the electron is deflected from the straight trajectory, but the energy of the electron remains constant. From Eq. 5.14, the Lorentz force on the incident electron being parallel to the x direction is expressed as

$$\frac{dP_x}{dt} = m\frac{d^2x}{dt^2} = evI_y \tag{5.15}$$

where m is the electron mass at velocity v,

$$m = m_e \Big/ \left(1 - \frac{v^2}{c^2}\right)^{\frac{1}{2}} \tag{5.16}$$

with m_e being the rest mass. In Eq. 5.15, $\frac{dP_x}{dt}$ is the x component of the Lorentz force, giving $P_x = m\frac{dx}{dt} = et_zI_y$, as the velocity of the electron can be treated as a constant. For the magnitude

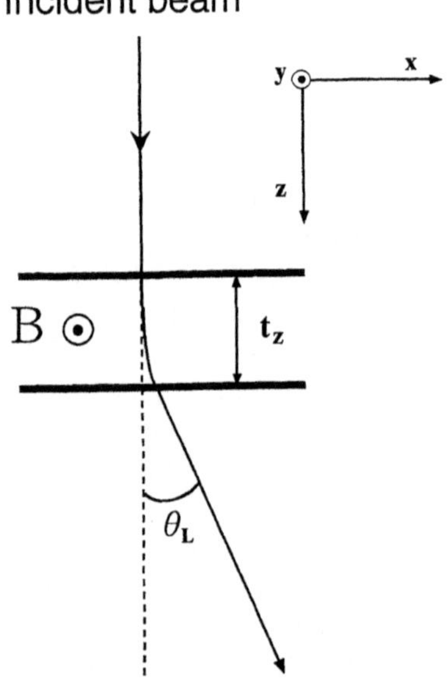

Fig. 5.9. Trajectory of an electron passing through a magnetized specimen

of magnetization given by $I_y = I_s$ and $I_x = I_z = 0$, the deflection angle (θ_L) is given by

$$\theta_L = \frac{dx}{dz} = \frac{dx}{dt} \cdot \frac{dt}{dz} = \frac{et_z}{mv} I_y \tag{5.17}$$

The deflection angle by the Lorentz force is rather small comparing with Bragg angles. For instance, the deflection angle in $Nd_2Fe_{14}B$ with thickness 100 nm and I of 1.6 T is 3×10^{-5} rad for 1250 keV electrons.

5.2.2 Fresnel Method (Defocusing Method)

The simplest domain formation is one consisting of 180° domains in a uniaxial material. Here we assume that the 180° domains are magnetized perpendicular to the plane of the diagram and in opposite directions in alternate domains (Fig. 5.10a). According to Fleming's left-hand rule, the electron beam is deflected in opposite directions in adjacent domains, resulting in a deficiency of electrons and an excess of electrons in the region below the lower specimen surface. If we observe the specimen in the overfocused condition, increases and decreases of the electron intensity are observed periodically at positions of the domain walls, as shown in Fig. 5.10b. On the other hand, if we observe the specimen in the under-focused condition, the reversed image contrast appears, as shown in Fig 5.10c. A typical example of Lorentz microscopic images of an as-sintered Sm-Co magnet observed by the Fresnel method is shown in Fig. 5.11. With the just-focused condition, there is no definite image contrast, whereas in the defocused condition white or black lines appear at positions on the magnetic domain walls depending on the sign of the defocusing. The width of the white or black lines depends directly on the magnitude of the defocused condition. Taking into account the experimental condition, such as carefully setting the focus, the wall width can be evaluated quantitatively, as shown below.

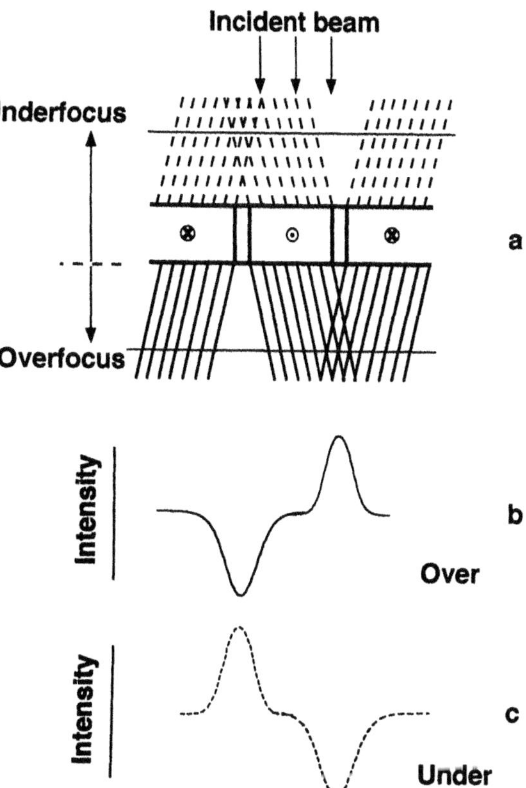

Fig. 5.10. **a** Fresnel method. Intensity distributions for the overfocused condition **b** and the underfocused condition **c**

Figure 5.12 shows a series of Lorentz microscope images observed by the Fresnel mode in an as-sintered Sm-Co magnet. The images were observed systematically by changing the defocus value. The wall widths were measured from the intensity profiles of divergent wall images recorded on imaging plates. These intensity profiles were obtained in the narrowest areas of divergent wall images to avoid overestimating the wall width because of the inclination of the domain wall against the electron beam. The wall widths plotted against the defocus values are on a straight line, as shown in Fig. 5.13; and eventually the domain wall width of an as-sintered Sm-Co magnet was evaluated to be 10 nm [17].

Figure 5.14 shows a comparison of the intensity distributions of the domain wall images in an as-sintered Sm-Co magnet observed by the Fresnel mode with the calculated intensity distributions. In Fig. 5.14, the squares and solid lines indicate the observed and calculated intensity distributions at the domain wall, respectively. Consider the spin distribution across the wall region for the minimum total energy resulting from exchange and either anisotropy or magnetostatic energy [18]; the equation of the theoretical intensity distribution of the domain wall images [19, 20] is derived as

$$I(U)/I = |1 \pm R\,\mathrm{sech}^2 u|^{-1}, U = u \pm R \tanh u \tag{5.18}$$

where u is x / δ (2δ is the wall width); U is X / δ; R is $\Delta f \theta_L / \delta$; x and X are the coordinates normal to the wall direction in the specimen and image planes, respectively; $I(U)$ is the electron beam intensity in the image plane; and I is the uniform illuminating intensity. The plus and minus signs in

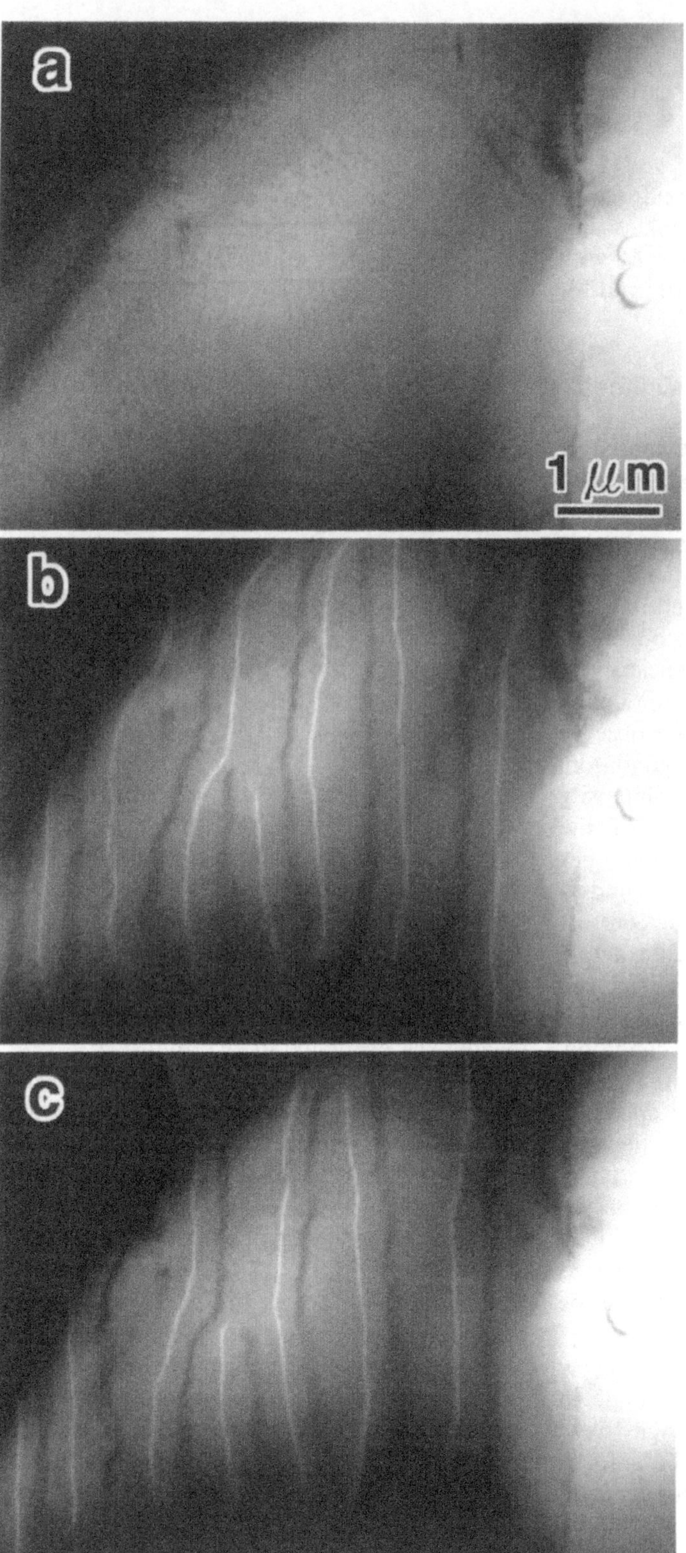

Fig. 5.11. Lorentz microscope images of an as-sintered Sm-Co magnet observed by the Fresnel mode under three conditions. **a** Just-focused. **b** Overfocused. **c** Underfocused

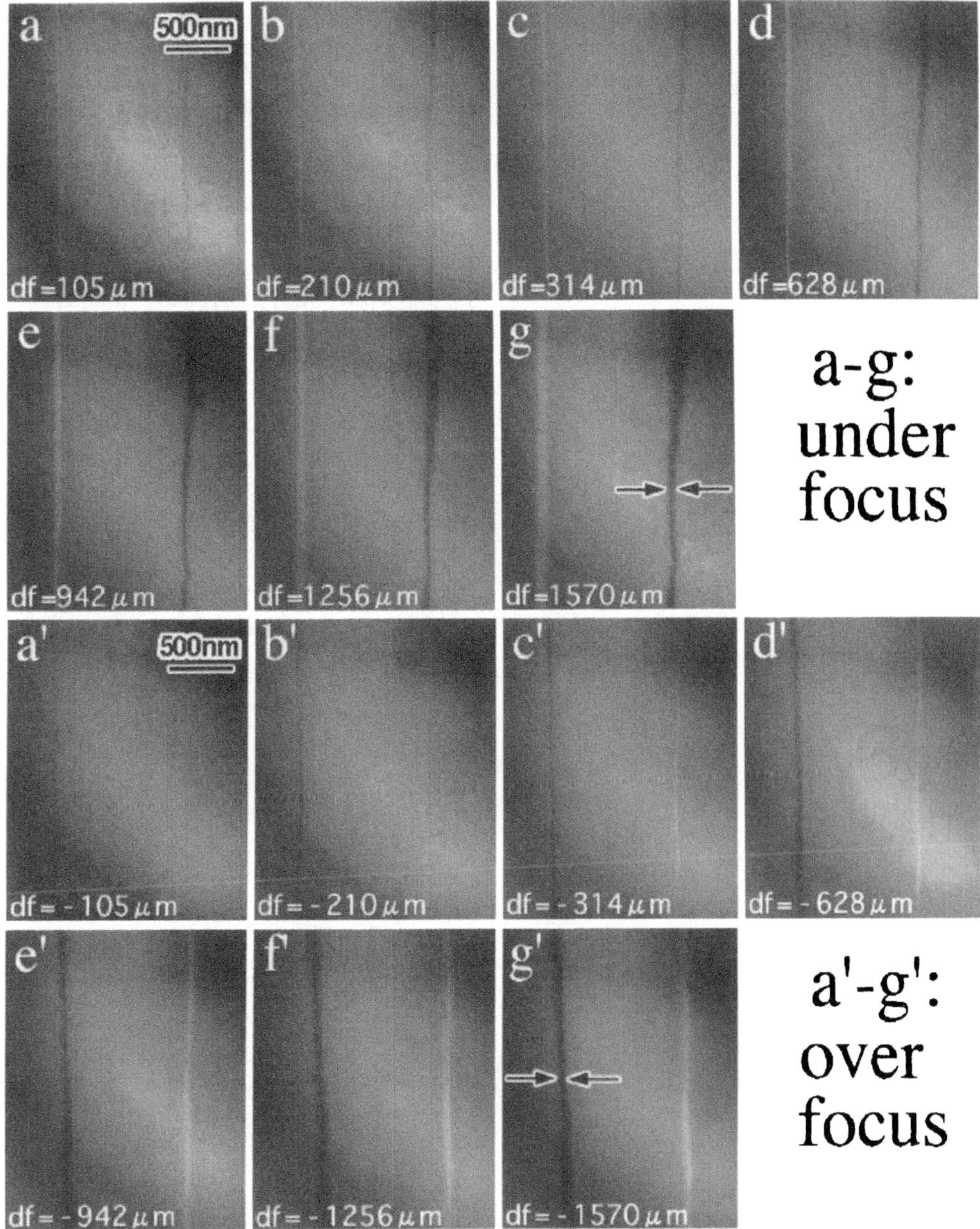

Fig. 5.12. Series of Lorentz microscope images of an as-sintered Sm-Co magnet for the focus change

Eq. 5.18 refer to the divergent and convergent cases, respectively. Equation 5.18 indicates that the width of white lines in the convergent images is significantly narrower than that of black lines in the divergent images. As shown in Fig. 5.14, the experimental intensity profiles measured with the imaging plate at a defocus value of 0.63 mm were best fitted with the calculated intensity profiles of the divergent and convergent cases at an R value of 0.3 [21].

5.2.3 Foucault Method (In-focus Method)

Figure 5.15 shows the principle of the Foucault method. In the diffraction pattern obtained from an area containing a domain wall separating two magnetic domains, such as shown in Fig. 5.15, the diffraction spot splits into two with an angular separation of $2\theta_L$. This splitting can be observed directly on diffraction patterns in both the transmitted and diffracted beams. In Fig. 5.16a, an elec-

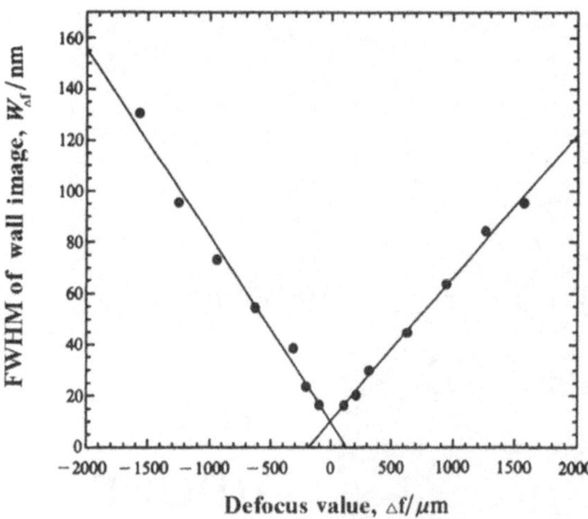

Fig. 5.13. Wall width measured as a function of the defocus value. *FWHM*, full width at half maximum

tron diffraction pattern obtained from a sintered $Nd_2Fe_{14}B$ magnet shows the split in each spot. The separation of the spots is about 5×10^{-5} rad, which is consistent with the estimation noted under Eq. 5.17. By displacing the objective aperture in the back focal plane, it is possible to exclude one of the two transmitted beams. As a result, alternate magnetic domains appear to be dark and bright regions periodically, as shown in Fig. 5.15. Figure 5.16b,c shows a typical example of the magnetic domains of an $Nd_2Fe_{14}B$ magnet observed by the Foucault mode.

Figure 5.17a,b shows Lorentz microscope images of step-aged Sm-Co magnets observed with the Fresnel mode and the Foucault mode, respectively. Whereas only the serrated domain walls indicating the pinning at the cell boundaries

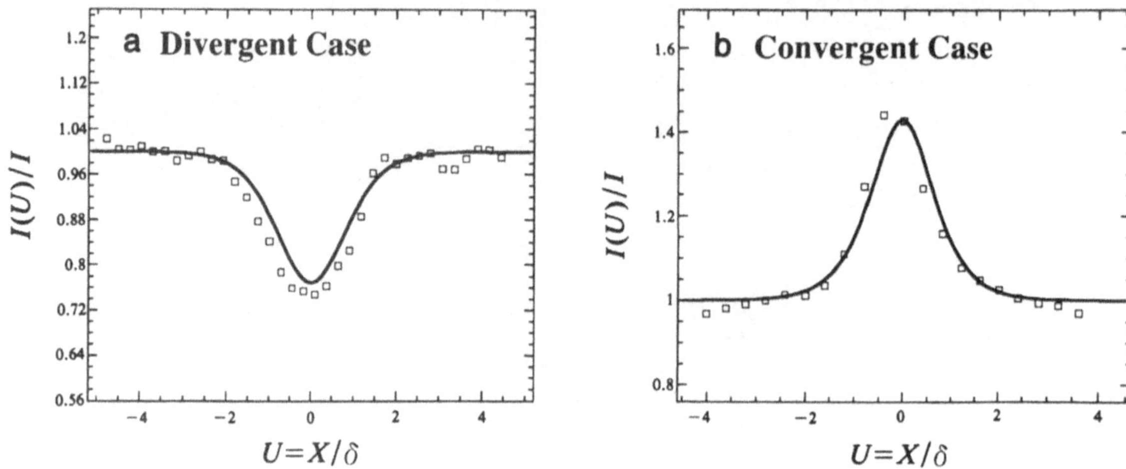

Fig. 5.14. Comparison of intensity distributions of the domain wall images in an as-sintered Sm-Co magnet with the calculated intensity distributions

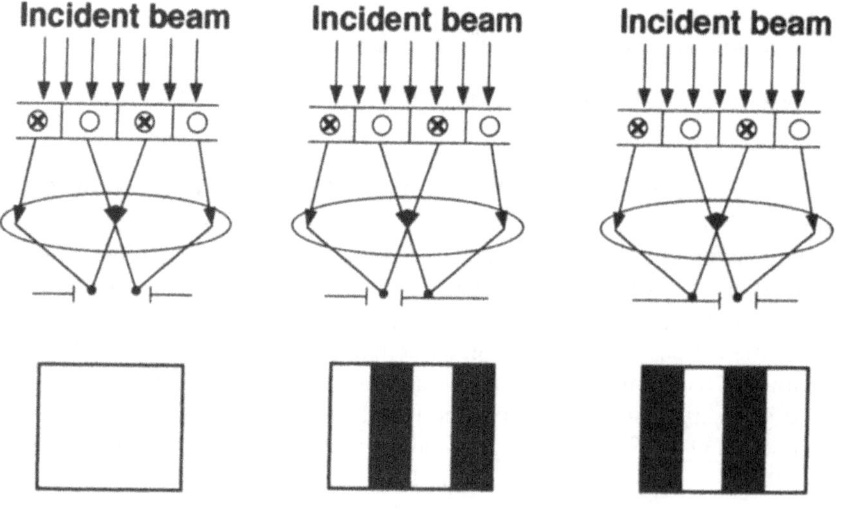

Fig. 5.15. Principle of the Foucault mode

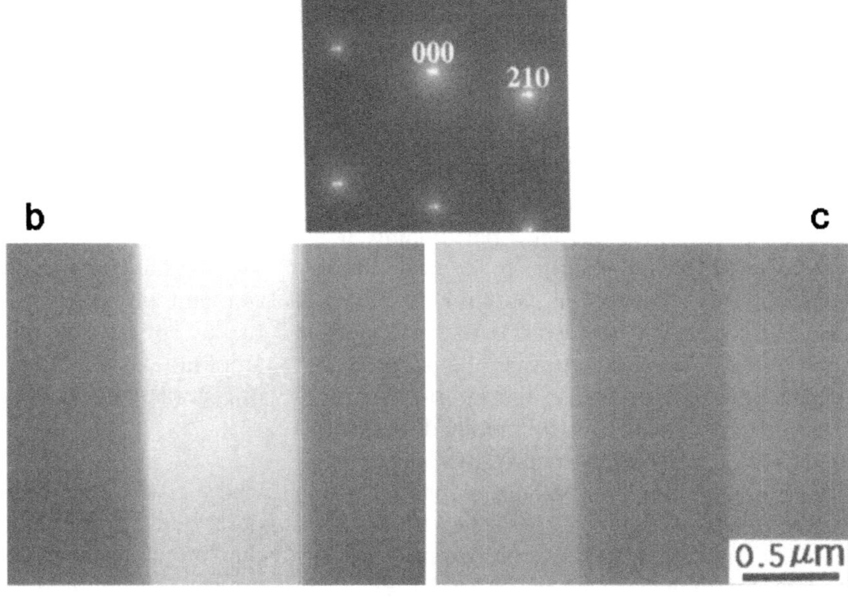

Fig. 5.16. **a** Electron diffraction pattern obtained from a sintered $Nd_2Fe_{14}B$ magnet showing the split in each spot. **b**, **c** Magnetic domains observed with one of the splitting transmitted beams

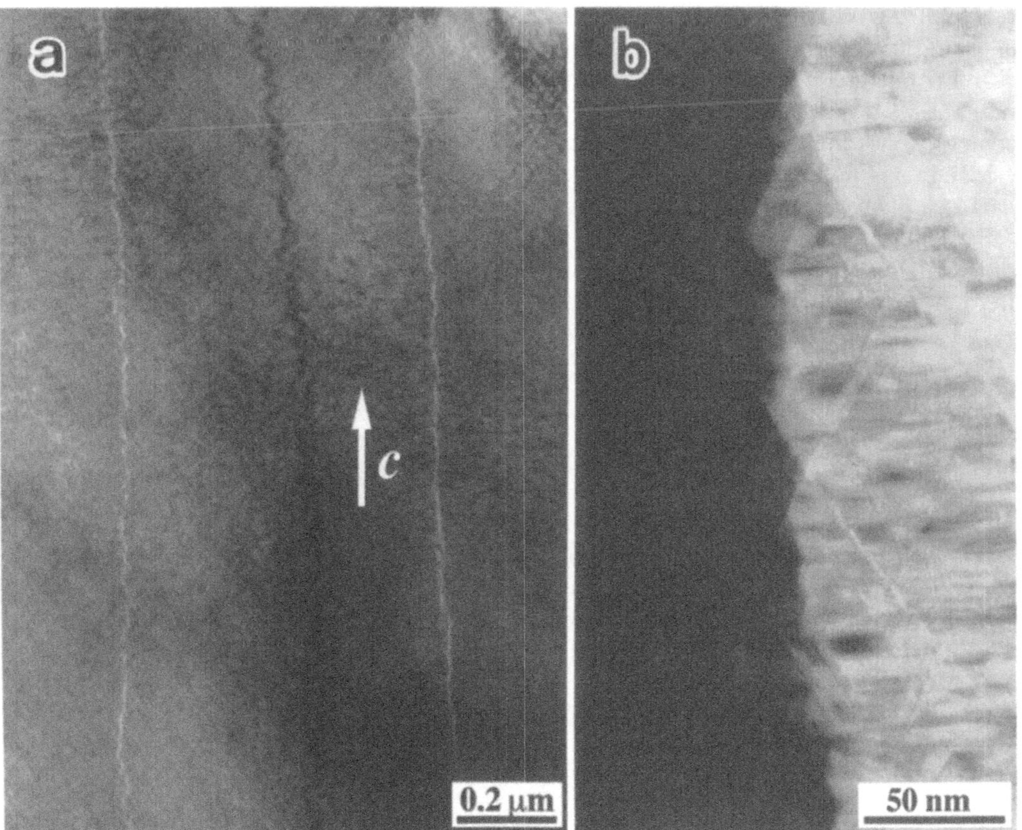

Fig. 5.17. Lorentz microscope images of a step-aged Sm-Co magnet observed with the Fresnel mode **a** and the Foucault mode **b**. *c*, the [001] direction

are observed in Fig. 5.17a, the Lorentz microscope image observed by the Foucault method in Fig. 5.17b clarifies not only the magnetic domain but also the microstructure consisting of so-called 1:5 H and 2:17 R phases where the 1:5 H phases of the cell boundaries act as the attractive pinning center for magnetic domain walls in a step-aged magnet. It can be also seen that domain wall pinning is greatly increased with an increase in the domain wall energy gradient between the 1:5 H and 2:17 R phases owing to the chemical partitioning of the additives of Cu and Fe atoms by step-aging [22].

A modified observation mode for the Foucault method has been developed [23]. With this method half of the diffraction pattern is eliminated with a phase-shifting aperture, so the interference contrast in the domain can be obtained. Apertures themselves are chosen to be of the opaque half-plane, phase-shifting half-plane, or phase-shifting small hole kind. The method is called the *coherent Foucault method*. If the edge of the phase-shifting aperture is put in the center of a transmitted beam, interference contrast is observed in all directions. The coherent Foucault image is obtained in real time without any postprocess, so the method is useful for in situ experiments [24]. It is also noted that Lorentz microscopy can be carried out using scanning TEMs. Like electron holography (presented in the following section), differential phase contrast (DPC) Lorentz microscopy introduced by Chapman et al. [25] provides detailed magnetization distribution in magnetic materials. The principle of DPC Lorentz STEM can be found in the literature [25, 26].

5.3 Electron Holography

5.3.1 Principles

Among various electron microscopy techniques, *electron holography* provides a unique method for detecting the phase shift of the electron wave due to the magnetic field and electrical field. Here, the phase shift means that the phase change relative to the electron plane wave was traveling in a vacuum, $\exp[i(kz-\omega t)]$, where

$$k = \frac{2\pi}{\lambda}, \quad \omega = 2\pi\upsilon \tag{5.19}$$

where υ is the frequency.

Electron holography is carried out through a two-step imaging process. In the first step, a *hologram* is formed by superimposing an *object wave* on a *reference wave* using a biprism. In the second step, the phase shift is extracted from the hologram using the Fourier transform.

Figure 5.18 shows a geometric configuration for forming a hologram in an electron microscope. An electron beam emitted from a field emission tip is accelerated and then collimated to illuminate an object through a condenser lens system. An object is located in one-half of the object plane being illuminated with a collimated electron beam. Assuming that the object is illuminated by a plane wave of a unit amplitude having a wave vector parallel to the optical axis, the change of the scattering amplitude of the plane wave due to the object is, in general, described as

$$q(\vec{r}) = A(\vec{r})\exp[i\phi(\vec{r})] \tag{5.20}$$

where $A(\vec{r})$ and $\phi(\vec{r})$ are real functions and describe the amplitude change and the phase shift due to the object, respectively. Because specimens are usually thin films, in most cases the vector $\vec{r}$ is confined in the film plane. Owing to the voltage of a biprism, the objective wave and the reference wave are tilted by $-\frac{\alpha_h}{2}$ and $\frac{\alpha_h}{2}$, respectively.

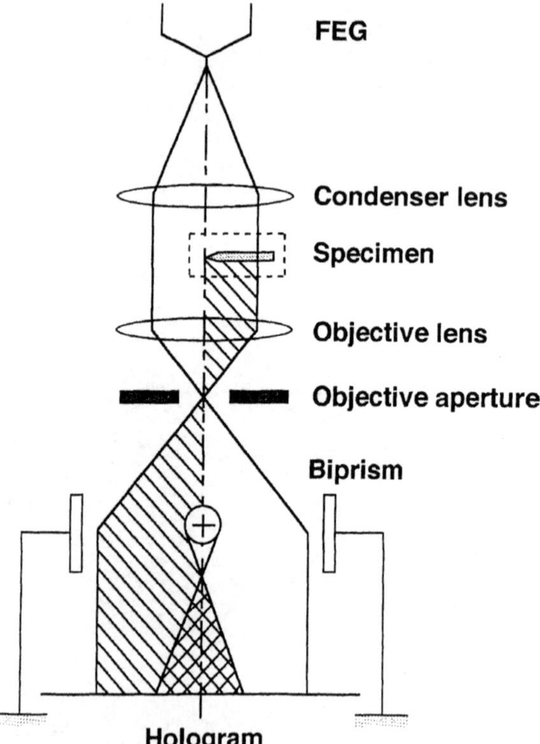

Fig. 5.18. Geometric configuration for forming a hologram in an electron microscope. *FEG*, field emission gun

Thus, the scattering amplitude resulting from the interference between the objective wave and reference wave is given by

$$g_h(\vec{r}) = A(\vec{r})\exp\left(-\pi i\frac{\alpha_h}{\lambda}x + \phi(\vec{r})\right) + \exp\left(\pi i\frac{\alpha_h}{\lambda}x\right) \tag{5.21}$$

The intensity of the hologram is given by

$$\begin{aligned} I_h(\vec{r}) &= |g_h(\vec{r})|^2 \\ &= 1 + A(\vec{r})^2 + 2A(\vec{r})\cos\left[2\pi\frac{\alpha_h}{\lambda}x - \phi(\vec{r})\right] \end{aligned} \tag{5.22}$$

According to Eq. 5.22, the period $\left(\frac{\lambda}{\alpha_h}\right)$ of the interference fringes in the hologram is modulated by the phase change $\phi(\vec{r})$ due to the object. By performing the Fourier transform ($\mathcal{F}$) of the hologram, one obtains

$$\begin{aligned} \mathcal{F}[I_h(\vec{r})] &= \delta(u) + \mathcal{F}\left[A(\vec{r})^2\right] \\ &+ \mathcal{F}\{A(\vec{r})\exp[-i\phi(\vec{r})]\} * \delta\left(u - \frac{\alpha_h}{\lambda}\right) \\ &+ \mathcal{F}\{A(\vec{r})\exp[i\phi(\vec{r})]\} * \delta\left(u + \frac{\alpha_h}{\lambda}\right) \end{aligned} \tag{5.23}$$

where the first and second terms are called "autocorrelation" (corresponding to the central area in Fig. 5.26c, below), and * is the convolution operation. The other two terms are called "sidebands." It is seen that the holographic information about the phase shift and the amplitude change is reserved at the sidebands in the third and fourth terms of the right-hand side of Eq. 5.23. By selecting the third term, shifting it by $-\frac{\alpha_h}{\lambda}$, and performing the inverse Fourier transform ($\mathcal{F}^{-1}$) on it, we obtain

$$\begin{aligned} &\mathcal{F}^{-1}(\mathcal{F}\{A(\vec{r})\exp[i\phi(\vec{r})]\} * \delta(u)) \\ &= A(\vec{r})\exp[i\phi(\vec{r})] \end{aligned} \tag{5.24}$$

Thus, we can obtain the phase shift and the amplitude change in the reconstructed image as digital data. In the following, the intensity of the reconstructed phase image $I_{ph}(\vec{r})$ is represented by cosine function, i.e.,

$$I_{ph}(\vec{r}) = \cos[\phi(\vec{r})] \tag{5.25}$$

If necessary, amplification of the phase can be done by multiplying the phase $\phi(\vec{r})$ by an integer n.

Now we consider the phase shift due to the electric potential φ and the vector potential $\vec{A}$ as follows. In general, the phase shift due to these potentials is given by

$$\phi(\vec{r}) = \frac{e}{\hbar}\oint(\varphi dt - \vec{A}d\vec{s}) \tag{5.26}$$

where e is the elementary electric charge; $\hbar$ is Planck's constant divided by 2π, and the integration goes over any closed circuit in space-time [27]. For the electrical potential only, the phase shift $\phi(\vec{r})$ is simplified as

$$\phi(x,y) = \sigma\int\varphi(x,y,z)\,dz \tag{5.27}$$

$$\sigma = \frac{2\pi}{\lambda V(1+\sqrt{1-\beta^2})}, \quad \beta = v/c \tag{5.28}$$

where V is the accelerating voltage; λ is the electron wave length; v is the electron velocity; c is the light velocity; and σ is the interaction constant, which depends on the accelerating voltage of an electron microscope. On the other hand, when t (time) is constant and the film is of uniform thickness as shown in Fig. 5.19, the phase difference between the two specimen positions is produced by the vector potential as given by

$$\phi(\vec{r}) = -\frac{e}{\hbar}\oint\vec{A}d\vec{s} \tag{5.29}$$

Here, we evaluate the phase difference of the electron wave at C and D due to the magnetic flux inside the specimen, where the electron wave has the same phase at A and B. The phase difference between C and D is given by

$$\begin{aligned} \phi(\vec{r}_C) - \phi(\vec{r}_D) &= -\frac{e}{\hbar}\int_A^C \vec{A}\cdot d\vec{s} + \frac{e}{\hbar}\int_B^D \vec{A}\cdot d\vec{s} \\ &= \frac{e}{\hbar}\oint_{ABDC} \vec{A}\cdot d\vec{s} \end{aligned} \tag{5.30}$$

From the definition of the vector potential

$$\vec{B} = \mathrm{rot}\vec{A} \tag{5.31}$$

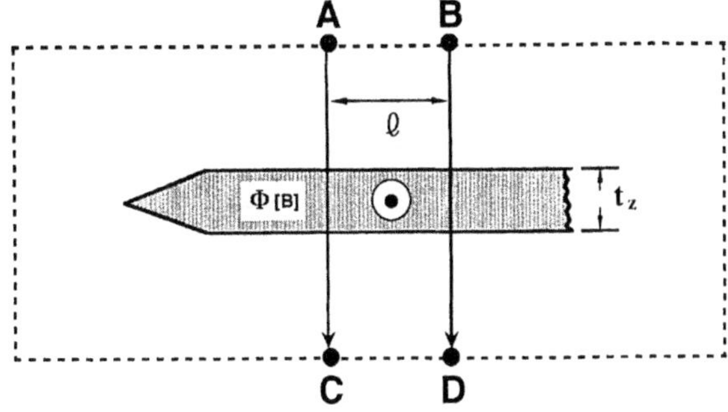

Fig. 5.19. Magnetic specimen in an electron microscope corresponding to the rectangular region indicated by dotted lines in Fig. 5.18

and using Stokes' theorem, the following relation is obtained.

$$\oint \vec{A} \cdot d\vec{s} = \iint \mathrm{rot}\vec{A} d\vec{S} = \iint B_n dS \qquad (5.32)$$

Thus, we find

$$\phi(\vec{r}_C) - \phi(\vec{r}_D) = \frac{e}{\hbar} \iint B_n dS = \frac{e}{\hbar} \Phi \qquad (5.33)$$

where Φ is the magnetic flux going through and being normal to the area ABDC. If the phase difference between $\vec{r}_C$ and $\vec{r}_D$ is 2π,

$$\phi(\vec{r}_C) - \phi(\vec{r}_D) = 2\pi \qquad (5.34)$$

From Eq. 5.33,

$$\frac{e}{\hbar} \Phi = 2\pi \qquad (5.35)$$

then

$$\Phi = \frac{h}{e} = 4.1 \times 10^{-15} \mathrm{wb} \qquad (5.36)$$

This is just twice the flux quantum $\left(\frac{h}{2e}\right)$. In this way, we have the relation between the magnetic flux inside the specimen and the width ℓ, which corresponds to the phase difference 2π, i.e.,

$$\Phi = \ell t_z B = \frac{h}{e} \qquad (5.37)$$

where we assumed a constant magnetic flux density B inside the specimen. Thus, the distance corresponding to the phase difference 2π is

$$\ell = \frac{h}{e t_z B} \qquad (5.38)$$

In the reconstructed phase images, such as those presented by cosine function, ℓ is the distance between the white lines or the black lines [28].

5.3.2 Practice of Electron Holography

5.3.2.1 Thickness Measurement

From Eq. 5.27, the thickness can be evaluated if the mean inner potential is known. The mean inner potential can be calculated from the structure factor, but the accuracy of the structure factor directly depends on the scattering factors of constituent elements. On the other hand, the mean inner potential can be determined experimentally if the shape of the specimen is known. As an example, the determination of the mean inner potential of amorphous SiO_2 is given below. Figure 5.20a shows an electron hologram of an amorphous SiO_2 particle. The interference fringes are on a part of the particle whose spherical shape is shown in a conventional TEM image (Fig. 5.20b). In the enlarged hologram of Fig. 5.20c, it can be seen that the inteference fringes shift at the particle edge where the thickness increases drastically. Figure 5.21 shows the phase shift evaluated from the interference fringes at lines X and Y. While the phase shift at the vacuum (X) is zero,

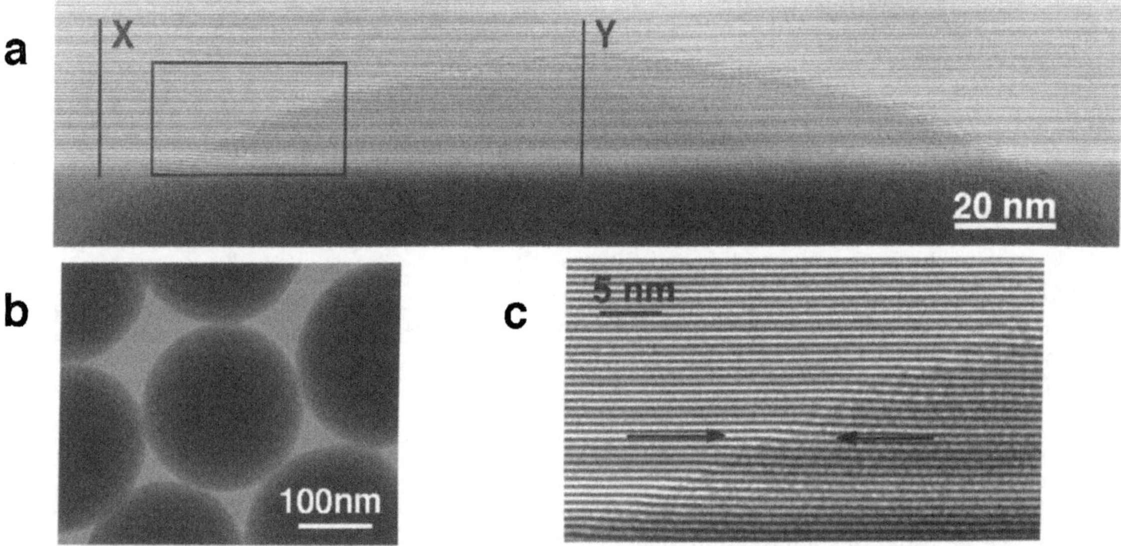

Fig. 5.20. **a** Electron hologram of an amorphous SiO_2 particle. **b** Conventional TEM image of the spherical SiO_2 particles. **c** Enlarged hologram of a rectangular region in **a**

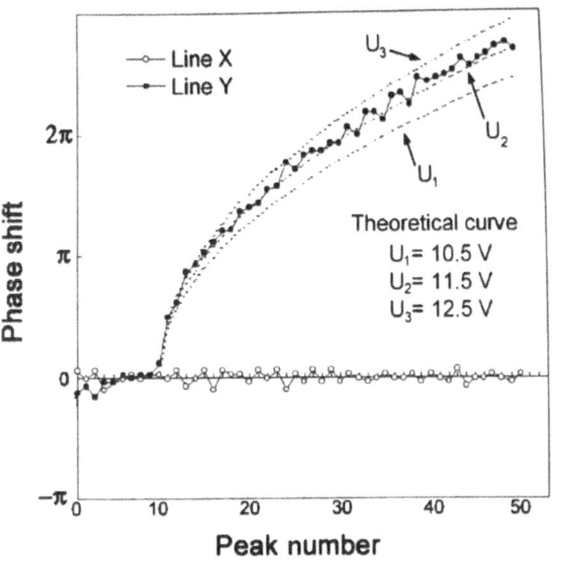

Fig. 5.21. Phase shift evaluated from the interference fringes at lines *X* and *Y*

that at the particle increases drastically from the particle edge. Because the particle has a spherical shape, the thickness of any part of the particle is easily estimated. Dotted lines correspond to the phase shift calculated by assuming the mean inner potential to be 10.5, 11.5, and 12.5 V. From a comparison of the experimental and calculated data, the mean inner potential of amorphous SiO_2 is evaluated to be 11.5 V [29]. Now with this mean inner potential, the thickness of amorphous SiO_2 can be evaluated from holograms without knowing its shape. On the other hand, with larger particles and higher electron intensity, a considerable charging effect is observed on the amorphous SiO_2 particles, which are insulators. The phenomenon is thought to result from the imbalance of the number of secondary electrons emitted and that of the incident electrons. Generally, a larger number of secondary electrons than incident electrons

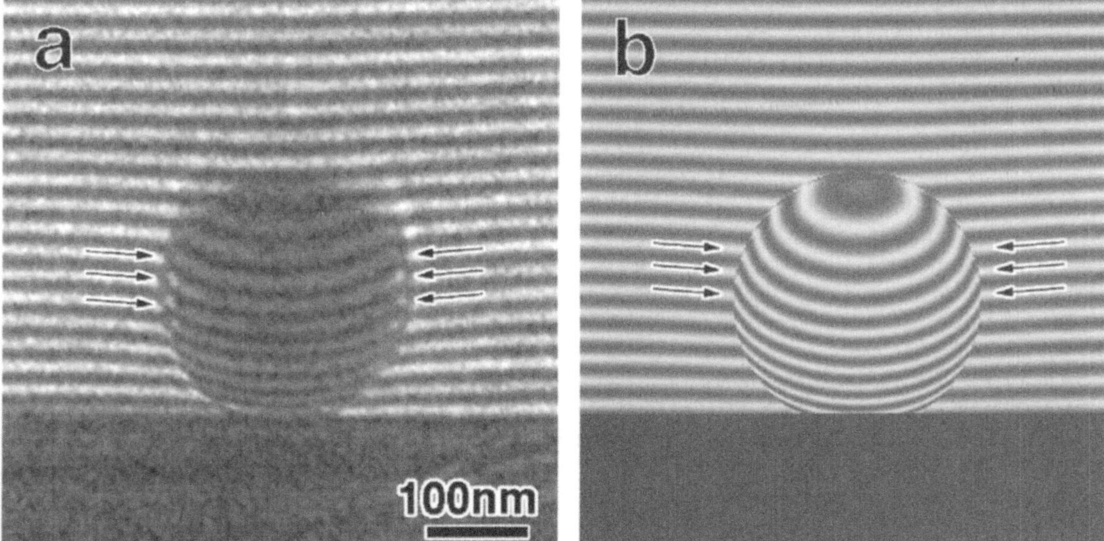

Fig. 5.22. Hologram of an amorphous SiO_2 particle of 250 nm diameter **a** and its simulation **b**. *Arrows* indicate the shift of the interference fringes near the particle surface due to the charging effect

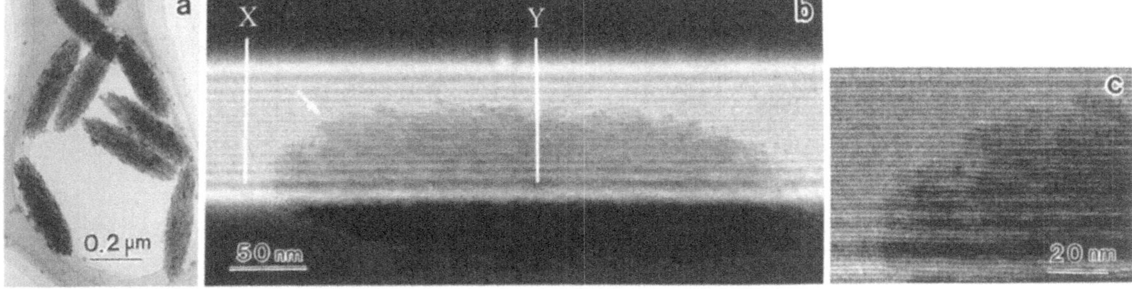

Fig. 5.23. **a** TEM image of hematite particles with a spindle shape. **b, c** Holograms of hematite particles and enlarged imaged of the region indicated by the *arrow* in **b**

leave the particle. Figure 5.22 shows a hologram of an amorphous SiO_2 particle of 250 nm diameter and its simulation. As indicated by arrows, the interference fringes shift near the particle surface owing to the charging effect.

If the mean inner potential is known, the thickness can be evaluated. Figure 5.23b,c shows holograms of hematite (α-Fe_2O_3) particles with a spindle shape (Fig. 5.23a). The particles are polycrystallin, consisting of small single crystals. Figure 5.24 shows the phase shift at lines X (vacuum) and Y (particle). It is seen that the phase shift at the particle does not increase monotonically. Fluctuation of the phase shift is attributed to the density fluctuation of single crystals and surface morphology. Furthermore, the density of the single crystals or the effective thickness could be evaluated with the known mean inner potential of hematite [30].

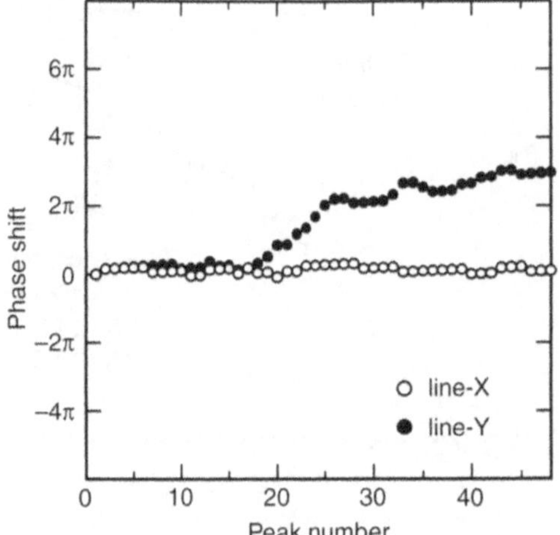

Fig. 5.24. Phase shifts at lines X (vacuum) and Y (particle) in Fig. 5.23b

Three-dimensional Information and Stereomicroscopy

In general, information obtained by electron microscopy is two-dimensional data projected along the incident electron beam. For example, bright-field, dark-field, and high-resolution microscopy as well as electron diffraction are utilized to obtain such data. On the other hand, three-dimensional information (e.g., specimen shape and a defect's position) are sometimes necessary to evaluate the dynamical diffraction effect and defect distribution, respectively. EELS (see Sect. 3.5.2), convergent beam electron diffraction (see Sect. 5.1.2), and electron holography are described in this book as techniques to measure crystal thickness. In addition, stereomicroscopy is one of the most powerful methods for obtaining three-dimensional information.

We observe a three-dimensional object by using both eyes. In other words, we obtain information about the distance by utilizing two images observed from different directions. This is the principle of stereomicroscopy. The method for measuring the height difference between the two points in a specimen by stereomicroscopy can be explained briefly. First, the rotational axis is set to be the x-axis, as shown in Fig. 5.25, and two images, one of which is tilted, are observed. Then the distance h from point P to the plane (xy plane), which includes the rotational axis and is perpendicular to the incident electron beam, is measured. When position P shifts to P' by tilting the specimen by θ_s, its distance from the rotational axis changes from l_p to $l_{p'}$. Then, h_p is given as

$$h_p = \frac{l_{p'}}{\sin\theta_s} - \frac{l_p}{\tan\theta_s} \tag{5.39}$$

If there is another signal point Q and the distance l_q changes to $l_{q'}$ with the tilt by θ_s, the height h_q from the xy-plane is given as

$$h_q = \frac{l_{q'}}{\sin\theta_s} - \frac{l_q}{\tan\theta_s} \tag{5.40}$$

From Eqs. 5.39 and 5.40, the height difference between the two positions P and Q is given by

$$\begin{aligned} |h_q - h_p| &= \left| \frac{l_{q'} - l_{p'}}{\sin\theta_s} - \frac{l_q - l_p}{\tan\theta_s} \right| \\ &= \left| \frac{l_{q'-p'}}{\sin\theta_s} - \frac{l_{q-p}}{\tan\theta_s} \right| \end{aligned} \tag{5.41}$$

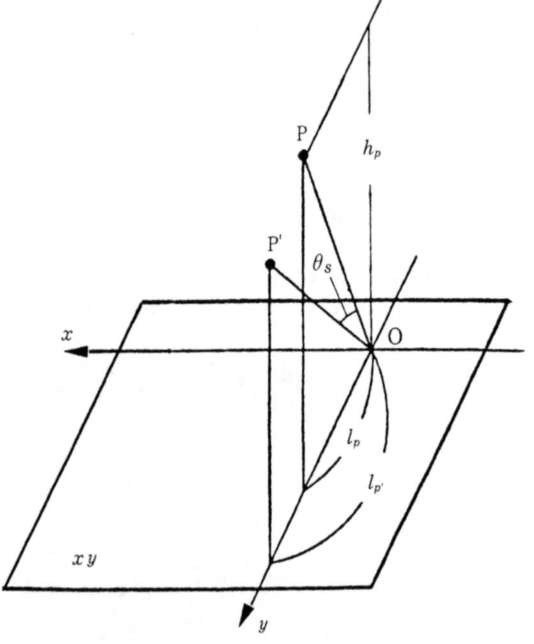

Fig. 5.25. Principle of stereomicroscopy

where l_{q-p} and $l_{q'-p'}$ are the y components of the distance between P and Q before and after the tilt, respectively.

Thus, the height difference between two points can be measured by stereomicroscopy if one knows the direction of the tilt axis accurately. If the tilt angle is small, the error mainly results from the accuracy of the evaluation of θ_s, whereas an out-of-focus image appears when the tilt angle is large.

5.3.2.2 Domain Structure Analysis

Here we present the analysis of domain structure of soft magnetic materials $Fe_{73.5}Cu_1Nb_3Si_{13.5}B_9$. The information obtained by electron holography is compared with that obtained by Lorentz microscopy [31]. Figure 5.26a shows a Lorentz microscope image of the as-quenched specimen. The Lorentz microscope image observed with the Fresnel mode shows magnetic domain boundaries as white and black bands (indicated by W1 and W2). An electron hologram of the same area as in Fig. 5.26a is shown in Fig. 5.26b. Because of the strong magnetic field of this material, the interference fringes curve from place to place. Figure 5.26c shows a digital diffractogram obtained from the electron hologram of Fig. 5.26b. The bright regions in the upper (circled) and lower parts of Fig. 5.26c correspond to the third and fourth terms of Eq. 5.23, respectively. After selecting the scattering amplitude of the circled region and translating it to the origin of the reciprocal space, an inverse Fourier transform is carried out to obtain a reconstructed phase image. When the

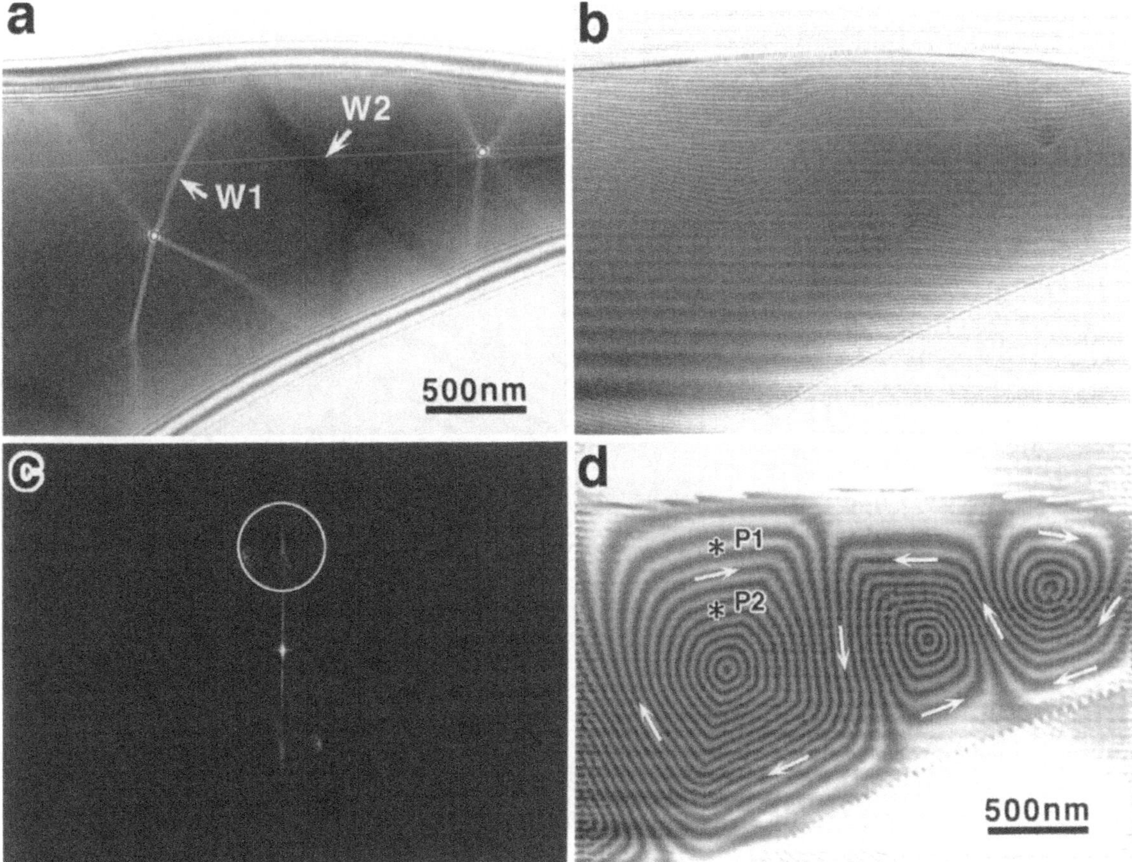

Fig. 5.26. **a** Lorentz microscope image of as-quenched $Fe_{73.5}Cu_1Nb_3Si_{13.5}B_9$. **b** Hologram of the same area as in **a**. **c** Digital diffractogram of **b**. **d** Reconstructed phase image

third term in Eq. 5.23 is picked up, the center of the sidebands to be shifted to the origin of the reciprocal space should be carefully selected. The digital diffractogram of the hologram without specimens can be utilized to find the center of the sidebands accurately. In the reconstructed phase image, represented by $\cos\phi(x,y)$ in Fig. 5.26d, the density and direction of the white lines indicate the density and direction (arrows) of the lines of magnetic flux, respectively. In Fig. 5.26d, smooth closure domains are clearly seen through the lines of magnetic flux. Note that the lines of magnetic flux are parallel to the specimen edges, thereby eliminating the surface magnetic charge. Also, the domain walls observed in Fig. 5.26a correspond to the boundaries in Fig. 5.26d, where the directions of the lines of magnetic flux change at about 90°. Note that the specimen thickness gradually increases from the edge to the middle of the specimen, and the spacing between the lines of magnetic flux become shorter with the increase in specimen thickness. If the magnetic flux density of a bulk specimen (1.28 T) is assumed, the crystal thickness can be simply estimated from Eq. 5.37. The thicknesses at positions P1 and P2 were estimated to be 36 and 54 nm, respectively. In the above evaluation, we neglected the phase shift of the incident electrons due to the inner potential; in other words, we assumed that the crystal thickness does not change appreciably at regions P1 and P2.

In Fig. 5.27a–c (top) the reconstructed phase images of the as-quenched specimen and the specimens annealed at 823 K and 973 K are shown, respectively. The reconstructed phase images obtained after tilting the specimens are shown at the bottom. By tilting the specimen, part of the residual magnetic field ($H_{//}$) can be introduced into the specimen film plane, as shown in Fig. 5.28. In Fig. 5.27a, it is interesting to note that the shape of the closure domain starts to change at a magnetic field of 8.3 A/m, which corresponds to the coercive force of the bulk specimen of 6.9 A/m. On the other hand, it is seen that the magnetic domain of the specimen annealed at 823 K, which shows the best property in this specimen as a soft mag-

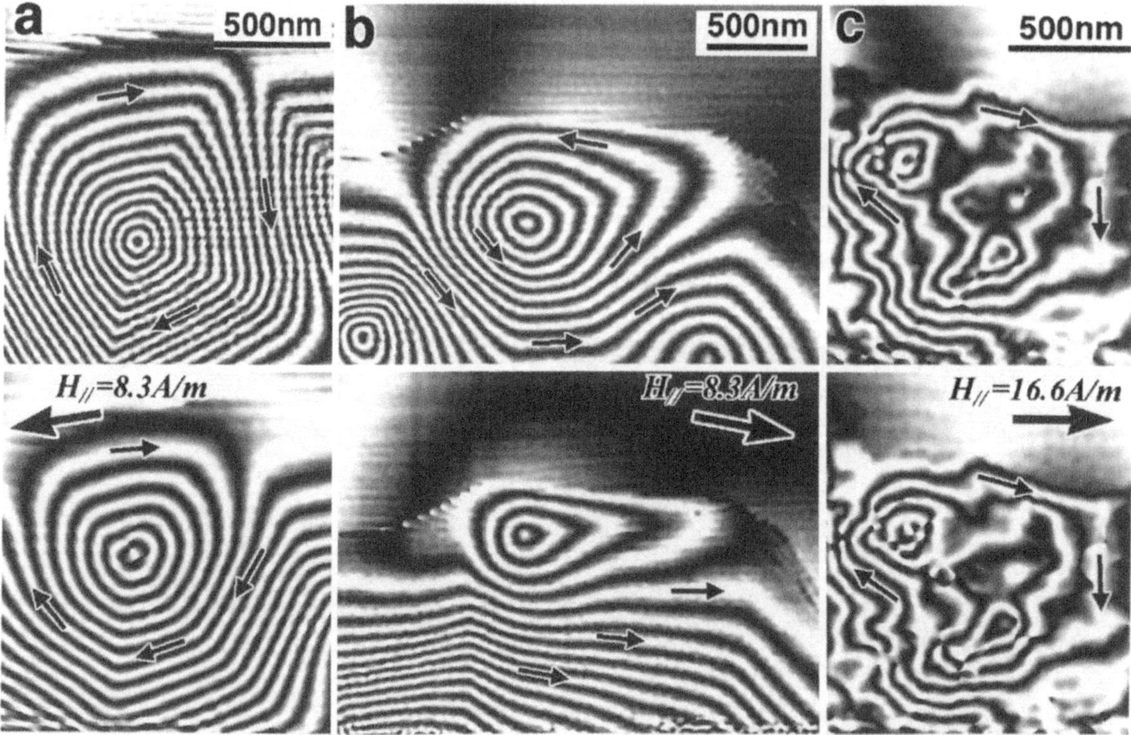

Fig. 5.27. Reconstructed phase images of as-quenched $Fe_{73.5}Cu_1Nb_3Si_{13.5}B_9$ **a** and specimens annealed at 823 K **b** and 973 K **c**. Reconstructed phase images obtained after tilting the specimens are shown in the bottom row. Big arrows indicate the directions of the magnetic field introduced

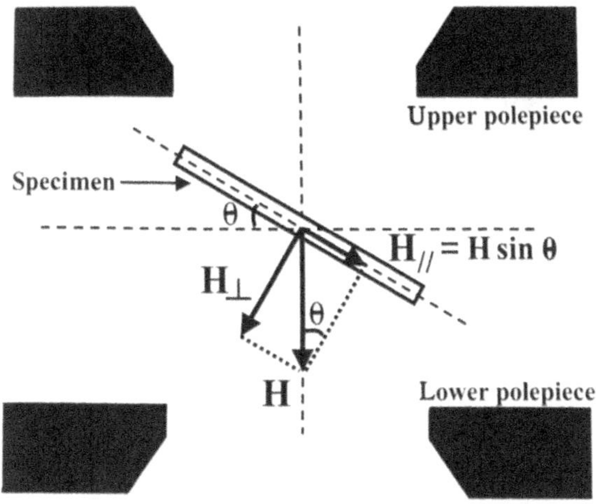

Fig. 5.28. Introduction of the residual magnetic field in the specimen film plane by tilting the specimen

netic material with low coercivity and high permeability, is more sensitive to the magnetic field, as seen in Fig. 5.27b. In the case of the specimen annealed at 973 K (Fig. 5.27c), the size of the magnetic domains becomes smaller, and the lines of magnetic flux deviate significantly from the monotonous line shape. The irregularity of the shape of the lines of magnetic flux is though to result from the inhomogeneous magnetization distribution due to the Fe-B compounds and the bcc Fe containing a small amount of Si and B. Different from Fig. 5.27a,b, the lines of magnetic flux do not change as much for the tilt. The difference directly indicates the strong pinning of magnetic domain walls due to the precipitates, resulting in a drastic increase in the coercive force and decreased permeability.

Effect of Phase Shift Due to Inner Potential and Magnetic Field on a Hologram and a Reconstructed Phase Image

To understand the image contrast of a hologram and a reconstructed phase image, it is important first to understand the direction of the shift of the interference fringes due to the inner potential (thickness effect) and the magnetic field.

Effect of Inner Potential on a Hologram. The holograms of a spherical SiO_2 particle are presented for the two conditions in Fig. 5.29, where the arrangements of the biprism and the particle are different for the two cases. In both cases the spacing of the interference fringes becomes wider when the thickness increases with increasing distance from the biprism position indicated in Fig. 5.29. Note that the positive phase shift in Eq. 5.22 results in wide spacing of the interference fringes; in other words, the interference fringes shift away from the biprism position. On the other hand, based on Fig. 5.29, when the thickness increases with a decrease in the distance from the biprism position, the spacing of the interference fringes narrows.

Effect of Magnetic Field on a Hologram and Reconstructed Phase Image. The change in the spacing of the interference fringes depends on the direction of the lines of magnetic flux, and it is important to know the relation between the spacing of the interference fringes and the direction of the lines of magnetic flux. It should be noted that the phase shift at position C relative to position D is positive when the direction of the line of magnetic flux is out-of-plane in Fig. 5.19. Thus, if there is a region where the direction of the lines of magnetic flux is as shown in Fig. 5.19, the spacing of interference fringes narrows with the decrease in

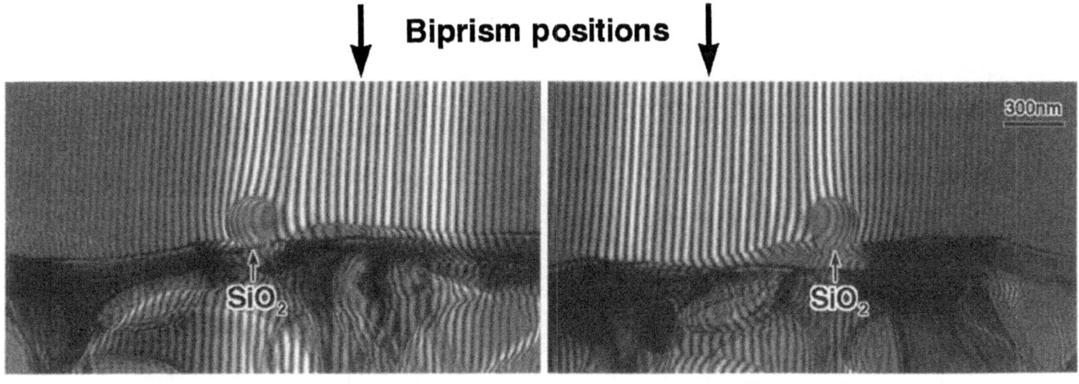

Fig. 5.29. Holograms of a spherical SiO_2 particle observed for two geometric conditions

distance from the position of the biprism (direction from D to C), as in the latter case of the effect due to the inner potential noted above.

In practice, the direction of the lines of magnetic flux can be determined from the change in the spacing of the interference fringes in the hologram, as follows (Fig. 5.30).

1. First, specify the direction of the incident electron beam. Here we assume the incident electron beam is incident on the picture (from front to back), being the same as the electron microscopic observation on a screen.

2. Find a region where the spacing of the interference fringes becomes narrower or wider with decreasing distance from the biprism. In the reconstructed phase image, there are lines of magnetic flux that are nearly parallel to the biprism in such regions (A and B in Fig. 5.30b). It is also important to find such regions where the thickness change is thought to be rather small, as the phase shift due to the inner potential should be avoided.

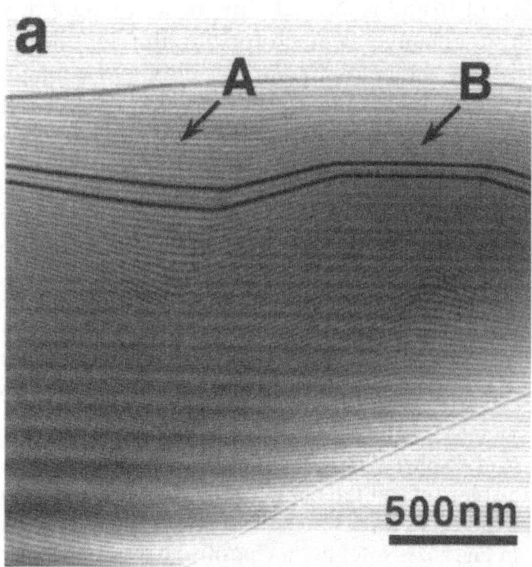

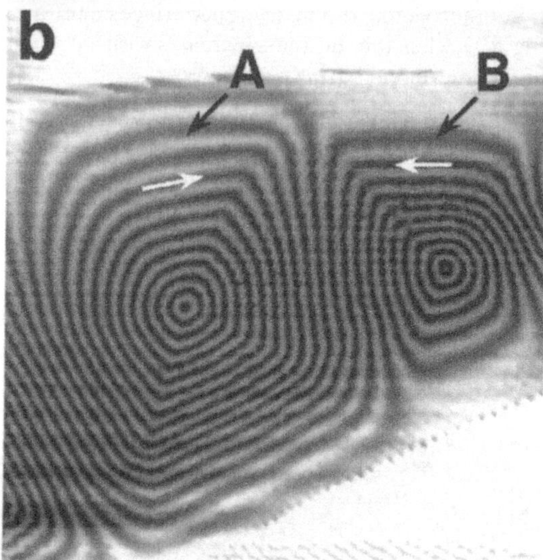

Fig. 5.30. **a** Enlarged hologram corresponding to a part of Fig. 5.26b. Two of the interference fringes are traced with black lines clarifying the shift of interference fringes. **b** Reconstructed phase image obtained from **a**. The position of the biprism is above the specimen (upper parts of the figures)

3. Apply the geometric configuration of Fig. 5.19 to this region. If the spacing of the interference fringes narrows with a decrease in the distance from the position of the biprism, being the same as the direction from D to C in Fig. 5.19, the direction of the lines of magnetic flux is the same as in Fig. 5.19. In contrast, if their spacing widens with a decrease in the distance from the position of the biprism, the direction of the lines of magnetic flux is opposite to that of Fig. 5.19. The former corresponds to region B in Fig. 5.30a and the latter to region A.

Note that the change of the phase shift can be easily estimated by representing the reconstructed phase image with $\phi(\vec{r})$ itself instead of the cosine function (Eq. 5.25).

Also determining the direction of the lines of magnetic flux is easier if one can take Lorentz micro-

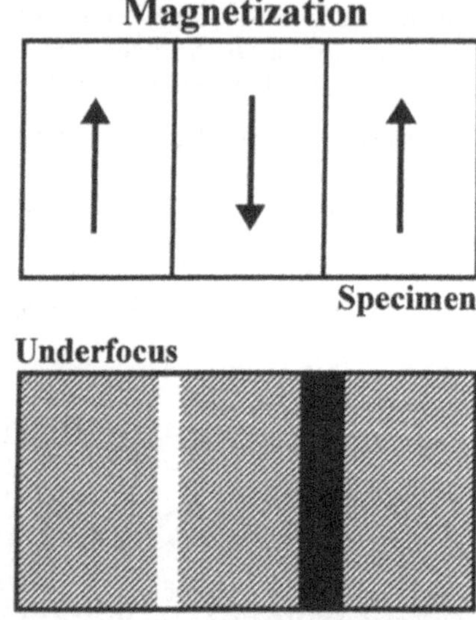

Fig. 5.31. Relation between the magnetization distribution in a specimen and the image contrast observed with the underfocused condition by the Fresnel mode. Electron beam is assumed to be incident on the plane (from front to back)

scope images showing domain boundaries at the same region. Because the image contrast of the domain boundary can be drawn as shown in Fig. 5.31, referring to Fig. 5.10 the direction of the line of magnetic flux can be specified by referring to this contrast of the domain boundaries, as demonstrated in Fig. 5.26.

Effect of Inner Potential and Magnetic Field on a Reconstructed Phase Image. In general, the reconstructed phase image showing lines of magnetic flux inside the magnetic material is affected more or less by the inner potential or thickness change. However, except for the crystal edges and small particles, where the thickness change are drastic, the effect of the thickness change is not large compared with that of the magnetic field in general. On the other hand, if the direction of the lines of magnetic flux is out of the specimen plane, the magnetic flux density projected along the incident electron beam becomes smaller and the thickness effect more prominent. As a typical case, simulation of the reconstructed phase image for a wedge-shaped magnetic specimen with the magnetization perpendicular to the incident beam is shown in Fig. 5.32. The reconstructed phase image of Fig. 5.32d is produced taking into account the effects of both the inner potential (Fig. 5.32b) and the magnetic flux (Fig. 5.32c).

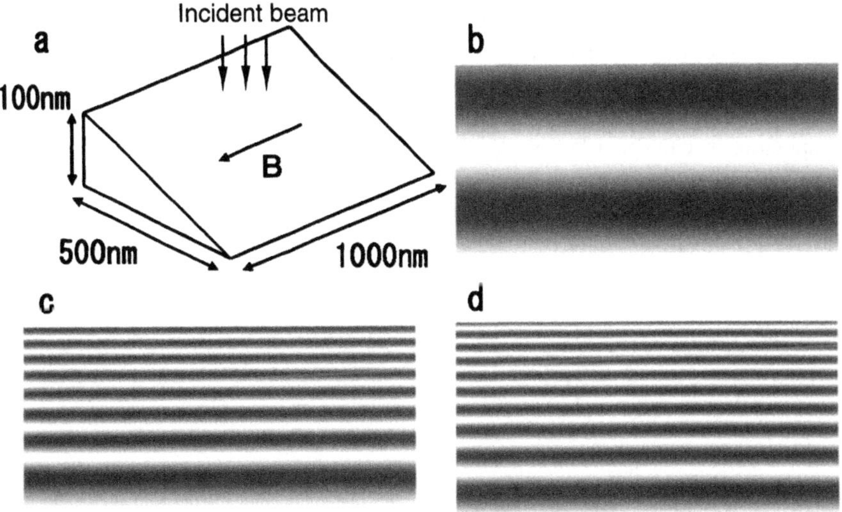

Fig. 5.32. **a** Magnetic wedge-shaped specimen. **b** Reconstructed phase image simulated taking into account the inner potential only. **c** Reconstructed phase image simulated taking into account the magnetic flux only. **d** Reconstructed phase image simulated taking into account both inner potential and magnetic flux. **b, c, d** The edge of the specimen is at the bottom of the figures; and the mean inner potential (17 V) and magnetic flux density (1.28 T) are assumed. The effect of the stray field is neglected

5.4 Scanning Electron Microscopy

An electron microscope in which a part of a specimen is illuminated with the electron beam and transmitted electrons are used for imaging is called a *transmission electron microscope* (TEM), and its observation mode is called transmission electron microscopy. A microscope in which a small electron probe is scanned on the specimen surface and the intensity of the resultant signals (e.g., secondary electrons) is displayed by synchronizing them with the probe position is called a *scanning electron microscope* (SEM), and its observation mode is called scanning electron microscopy. Usually, scanning electron microscopy has been carried out with secondary electrons or backscattered electrons at an accelerating voltage of 10–20 kV. SEM with lower accelerating voltage (about 1 kV) is also utilized. With an accelerating voltage higher than 100 kV, transmitted electrons from a thin specimen are utilized as signals in a *scanning transmission electron microscope* (STEM). Conventional SEM with relatively low accelerating voltage is deseribed in Section 5.4.1, and application of STEM, that is, the high-angle annular darkfield method, is explained in Section 5.4.2.

5.4.1 Principles and Application of SEM

Figure 5.33 shows the principles of an SEM. LaB_6 has been utilized so far as the electron gun. Nowadays a field emission gun (FEG) that produces a smaller probe is introduced. To reduce the probe size, a so-called in-lens type, where the specimen is inside the objective lens, is utilized. The resolution of an in-lens SEM is about 1 nm. Because a large specimen such as semiconductor wafers cannot be inserted in the in-lens type, a semi-in-lens SEM, where the wide space is provided for the specimen, is also utilized currently. The signals used are not only the secondary electrons but also backscattered electrons and cathode-ray luminescence. Unlike the thin specimens used for TEM images, bulk specimens can be observed in an SEM. When insulators are observed with an SEM, conductive materials such as carbon and Pt-Pd should be evaporated on the specimen surface or the specimen should be heated to avoid the charging effect. A low-vacuum SEM with a specimen chamber with low vacuum condition can be utilized as well. In this case, the charging effect of insulators is avoided, and the specimens containing a small amount of water can be also observed.

Figure 5.34a shows an SEM image of a platelet-type hematite (α-Fe_2O_3) particle. The SEM image is obtained mainly with secondary electrons, and the bright regions in the image correspond to such regions where many secondary electrons are emitted. When there are edge-shaped regions, the probability of the secondary electron emission here is high, and the regions appear bright. The phenomenon is the so-called edge effect. In Fig. 5.34a the edge parts A, C, and E are observed as brighter regions than the other edge parts (B, D, F), and so the former parts are considered protruded regions, and the latter are on the substrate, as shown in Fig. 5.34b.

Figure 5.35 shows an SEM image of silicon oxide film, with the silicon being sharply etched. The image was observed with an FEG-SEM at low accelerating voltage (7 kV). If higher accelerating

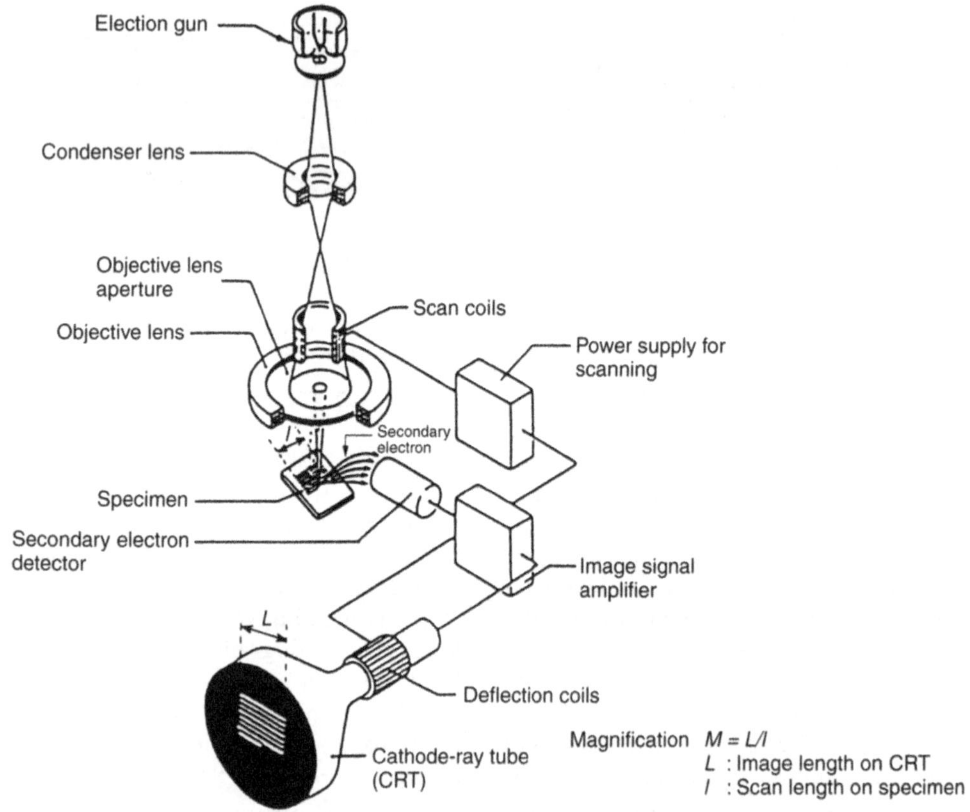

Fig. 5.33. Basic constitution of a scanning electron microscope (SEM)

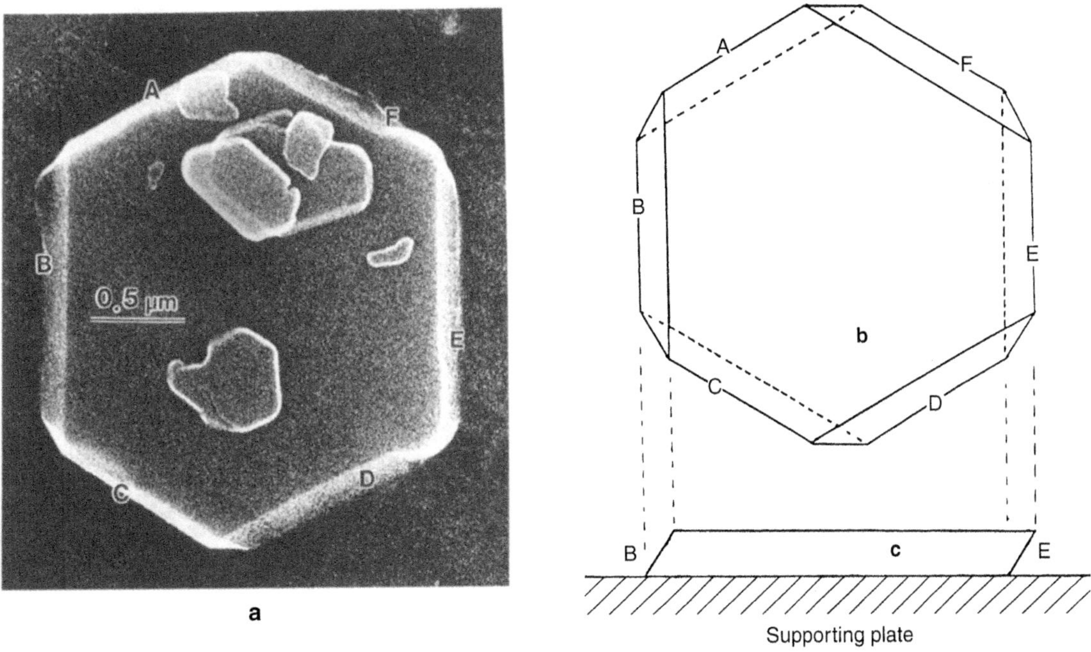

Fig. 5.34. **a** SEM image of a platelet-type hematite particle, showing its shape **b** and cross section **c**

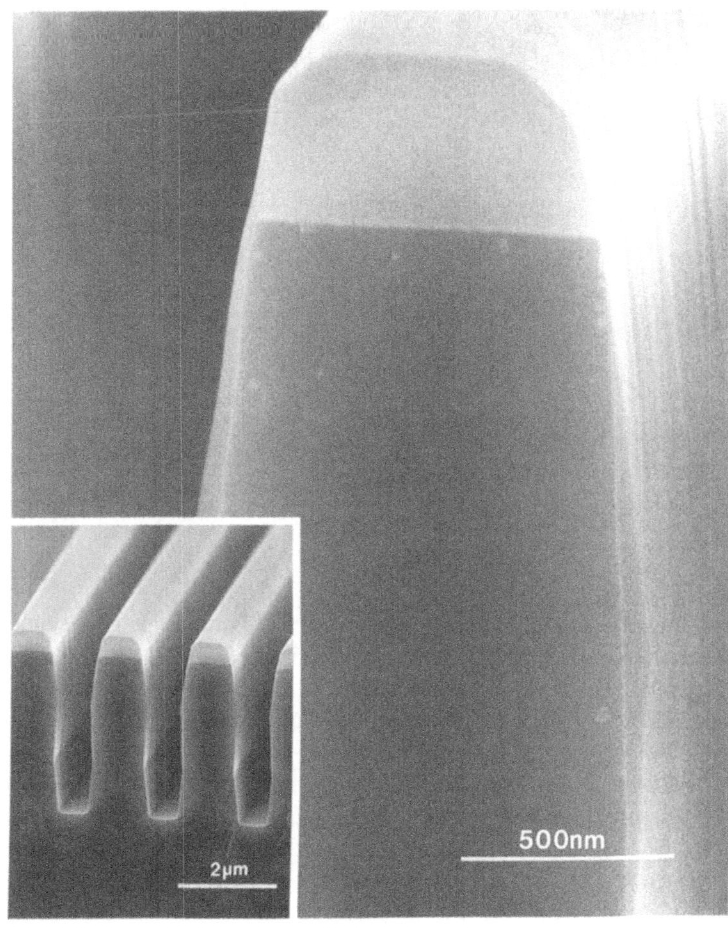

Fig. 5.35. SEM image of silicon oxide film on silicon prepared by dry etching. There is no coating. Accelerating voltage was 7 kV (FEG-SEM: JSM-890)

Fig. 5.36. Surface of paper observed with a low-vacuum SEM. Accelerating voltage was 15 kV (FEG-SEM: JSM-5300LV)

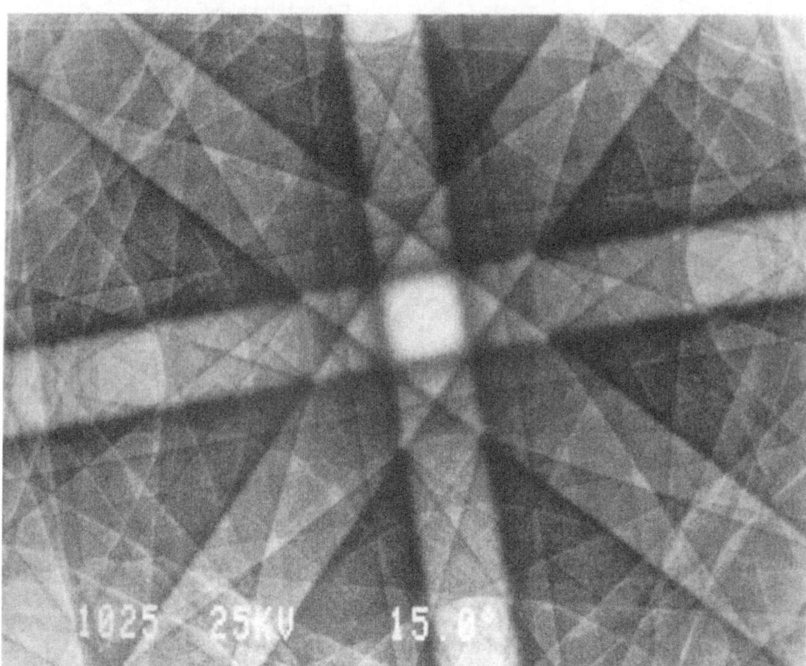

Fig. 5.37. Electron channeling pattern (ECP) of a silicon single crystal in the [100] direction. Accelerating voltage was 25 kV (JSM-880)

voltage is used, signals from a deeper region under the surface are enhanced, and so obtaining information about fine surface morphology is difficult. Accurate surface morphology is clearly imaged with lower accelerating voltage.

Figure 5.36 shows an SEM image of the surface of paper observed with a low-vacuum SEM. Because there is no surface coating, the original surface feature of paper is revealed. Thus, with a low-vacuum SEM the charging effect is avoided, and specimens containing a small amount of water or oil can be observed.

Figure 5.37 shows an electron channeling pattern (ECP) of a silicon single crystal. For observing an ECP, first the electron probe is stopped at a point of the specimen's surface, and backscattered electrons are detected by changing the incidence angle $\theta_{x,y}$ systematically. An ECP is then obtained by plotting the intensity of the backscattered electrons as functions of x and y, which correspond to $\theta_{x,y}$. An ECP is a diffraction pattern of backscattered electrons that corresponds to the Kikuchi pattern of TEM. An ECP is useful for investigating the orientation relation of the grains in a bulk specimen.

5.4.2 High-Angle Scattered Dark-field STEM

Elastically scattered electrons distribute at large scattering angles, whereas inelastically scattered electrons distribute at small scattering angles. Therefore, the elastically scattered electrons can be selectred by detecting the scatted electrons at large scattering angles. With this method the transmitted electrons, which are in the center of the diffraction pattern, are not detected. Therefore, the signal obtained by the beam scanning method with a STEM forms a dark-field STEM image. Since the distribution of the scattered electrons, except the Bragg reflection, has the rotational symmetry, an annular-shaped detector is widely used for high detection efficiency. This detecting mode is called a *high-angle annular dark-field* (HAADF) method.

Figure 5.38 shows the principle of the HAADF method. According to Pennycook et al. [32], the partial scattering cross section of the electrons distributed in the annular shaded area of Fig. 5.38 can be obtained by integrating the Rutherford scattering intensity from the scattering angle θ_1 to θ_2.

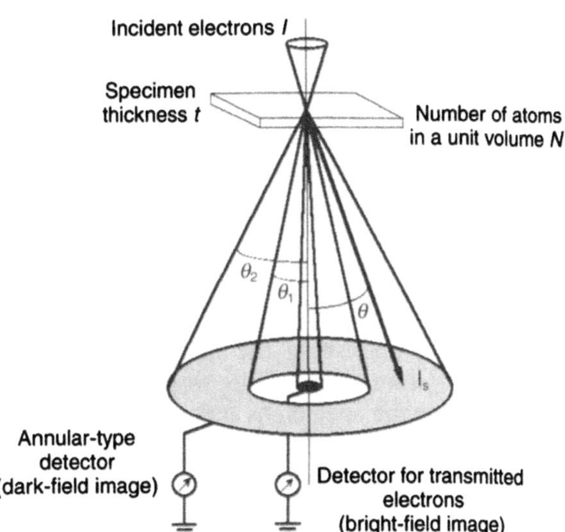

Fig. 5.38. Principle of the high-angle annular dark-field (HAADF) microscopy

$$\sigma_{\theta_1,\theta_2} = \left(\frac{m}{m_0}\right)^2 \frac{Z^2\lambda^4}{4\pi^3 a_0^2}\left(\frac{1}{\theta_1^2+\theta_0^2} - \frac{1}{\theta_2^2+\theta_0^2}\right) \quad (5.42)$$

where m is the electron mass; λ is the electron wavelength; m_0 is the electron rest mass; a_0 is the Bohr radius; Z is the atomic number; and θ_0 is the Born scattering angle.

When the number of atoms in a unit volume of the specimen is N, the scattering intensity I_s is written as

$$I_s = \sigma_{\theta_1,\theta_2} \cdot NtI \quad (5.43)$$

where I indicates the incident electron intensity.

From Eqs. 5.42 and 5.43, it is seen that the signal intensity of HAADF is proportional to the square of the atomic number Z. Therefore, the image contrast strongly depends on Z, and a HAADF image sometimes is called a *Z contrast image or* a Z^2 *contrast image*.

Because the image is formed with incoherent electrons, different from conventional HREM or bright-field STEM images, interpretation of the HAADF image is straightforward. The bright image contrast indicates heavy elements directly if the specimen thickness is uniform. The HAADF image has attracted much attention because of the ease of image interpretation and amplification of the image contrast electronically. On the other hand, care should be taken regarding the thickness change and diffraction contrast when the image contrast is interpreted quantitatively. Figure 5.39 shows an atomic-resolution HAADF image of a grain boundary in a semiconducting $SrTiO_3$ ceramics condenser [33]. The image was obtained with a 200 kV FE-STEM, which provides a beam diameter of less than 0.2 nm. Also, an attached HAADF detector collected the electrons at the scattering angle of 50–110 mrad. The columns of Sr with a large Z appear as bright dots, and the columns consisting of Ti and O are clearly seen. The structure of the model is shown at the bottom.

5.5 Specimen Preparation Techniques

Specimen preparation techniques for analytical electron microscopy are basically the same as those for conventional TEM. However, for EELS and energy dispersive X-ray spectroscopy (EDS), a specimen should be prepared to be as thin as possible to enhance the P/B ratio, reducing the background of the spectra. It should also be

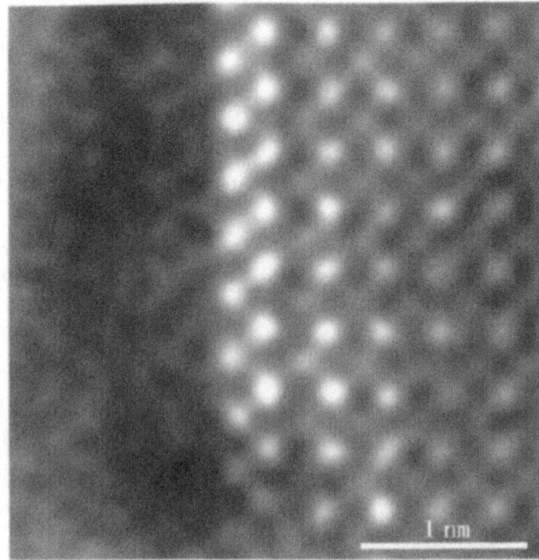

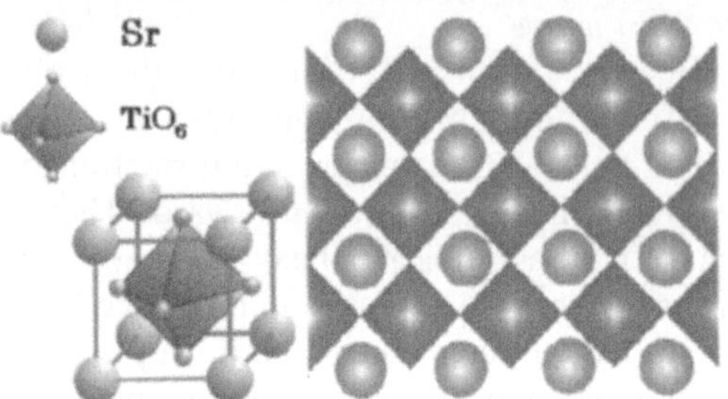

Fig. 5.39. Atomic-resolution HAADF image of a grain boundary in a semiconducting $SrTiO_3$ ceramic condenser. A 200 kV FE-STEM (JEM-2010F) was utilized. Structure model is shown at the bottom

noted that some specimen preparation techniques produce an impurity phase on the surface, which affects the accuracy of EELS and EDS. Typical specimen preparation techniques are outlined in the following sections, and some notice for analytical electron microscopy is noted.

5.5.1 Crushing Method

The crushing method is applied to oxides and ceramics. Although this method is the simplest among the specimen preparation techniques, thin regions (a few nanometers) with little contamination on the surface can be obtained with this method. It is especially useful for EELS. However, the application is limited to materials that tend to cleave.

A specimen is crushed usually with an agate mortar and pestle. The flakes obtained are suspended in an organic solvent such as butyl alcohol or acetone and are dispersed with supersonic waves or by simply stirring with a glass stick. Finally, the solvent containing the specimen flakes is dripped onto a microgrid on a filter paper. Figure 5.40 shows small particles of hematite (α-Fe_2O_3) obtained by stirring in butyl alcohol and dripped onto a microgrid. With analytical electron microscopy, attention should be paid to the background from the carbon film that supports specimens.

5.5.2 Electropolishing

Electropolishing is used mainly to prepare thin films of metals and alloys. First, a bulk specimen is sliced into thin plates about 0.3 mm in thickness by a fine cutter or a multiwire saw. A thin plate is further thinned mechanically down to about 0.1 mm. Electropolishing is performed in a specific (electrolyte) solution by supplying a direct current with a positive pole at the thin plate and a negative pole at a platinum plate or stainless steel plate. To avoid preferential polishing at the specimen edge, all the edge is covered with insulating paint. This is called the *window method*. The electropolishing is finished when there is a small hole at a central part of the plate with thin regions.

The so-called twin-jet polishing method is widely used nowadays. As shown in Fig. 5.41, the solution is jetted through two small nozzles onto the center of a specimen plate on both sides. Thus the specimen plate (0.1–0.2 mm thick and 3 mm in diameter) should be prepared in advance. This plate can be directly obtained with a disk puncher if the specimen is reasonably soft; otherwise a spark cutter should be used. In a conventional jet polishing machine, when a hole is formed in a central part of the disc, polishing is automatically stopped by the operation of a photocell. Solutions should be selected for each material, and polishing is performed at the appropriate temperature and voltage. Solutions to be used for specific materials and the appropriate operation conditions are noted elsewhere [1].

After electropolishing, the specimen should be washed as soon as possible with methanol or water. When the specimens are not washed correctly, contamination (e.g., an oxide layer) forms on the specimen surface. This surface layer produces a strong background in an energy-loss spectra. The existence of these oxide layers can be confirmed from an oxygen K-edge in EELS and oxygen-Kα line in EDS. In electron diffraction

Fig. 5.40. Hematite particles dispersed on a microgrid

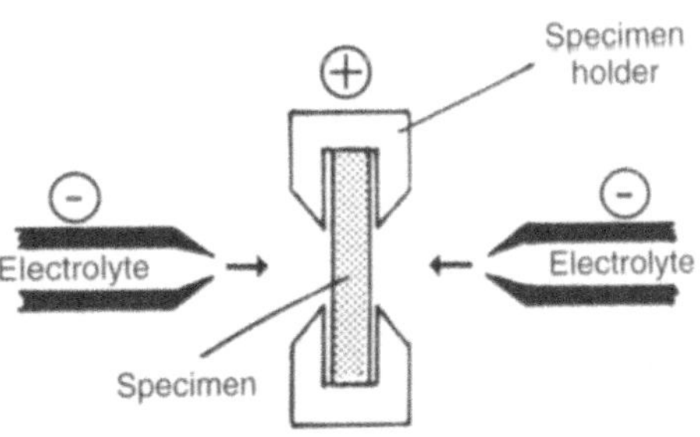

Fig. 5.41. Principle of the twin-jet polishing method

patterns, such oxide layers can be confirmed with *Debye-Scherrer rings* from a polycrystalline phase and a *halo ring* from an amorphous phase. It is sometimes useful to use ion milling to remove the oxide layers (see Sect. 5.5.5).

5.5.3 Chemical Polishing

Chemical polishing is frequently used for thinning semiconductors such as silicon. Thinning is performed chemically (i.e., by dipping a specimen in a specific solution). Like electropolishing, a thin plate (0.1–0.2 mm thick) should be prepared in advance. If a small dimple is formed at a central part of the plate by a dimple grinder, a hole can be made by etching around the center, keeping the edge of the specimen relatively thick. Appropriate solutions for specific materials are listed elsewhere [1].

As with electropolishing, if the specimen is not washed properly after chemical etching, contamination (e.g., an oxide layer) forms on the surface of the specimen that results in a background of energy-loss spectra. Ion milling is sometimes useful for removing this contamination.

5.5.4 Ultramicrotomy

Ultramicrotomy has been used to prepare thin sections of biological specimens and sometimes thin films of inorganic materials, which are not as difficult to cut. Specimens of thin films or powders are usually fixed in a resin and trimmed with a glass knife before being slicing by a diamond knife. This process is necessary so the specimens in the resin can be sliced easily with a diamond knife. Acrylic or epoxy resin is used for fixing specimens. When using acrylic resin, a gelatin capsule is used as a vessel; and the acrylic resin is easily sliced. Epoxy takes less time to solidify than acrylic, and it is rather strong for electron irradiation. In general, skill is needed to set the geometrical configurations of a diamond knife and a specimen appropriately after trimming and to slice a specimen into homogeneous thin sections.

Figure 5.42 illustrates the principle of ultramicrotomy. Each time the arm holding the specimen comes up and down, it steps forward; in this way the specimen is sliced with the diamond knife at the head of the boat filled with water. Sliced sections are on the water and are handled by a thin wooden stick with an eyelash to put thin sections onto a special grid covered with a collodion or carbon thin film. For analysis, attention should be paid to the background in EELS and EDS due to the resin and supporting film. The acrylic can be removed with chloroform after slicing. Figure 5.43 shows high-resolution electron microscope image of a section of a platelet-type hematite particle. The section corresponds to the cross section of the platelet-type particle shown in Fig. 5.34. The crystallographic planes of basal planes and side planes can be identified from the high-resolution images and electron diffraction patterns [34].

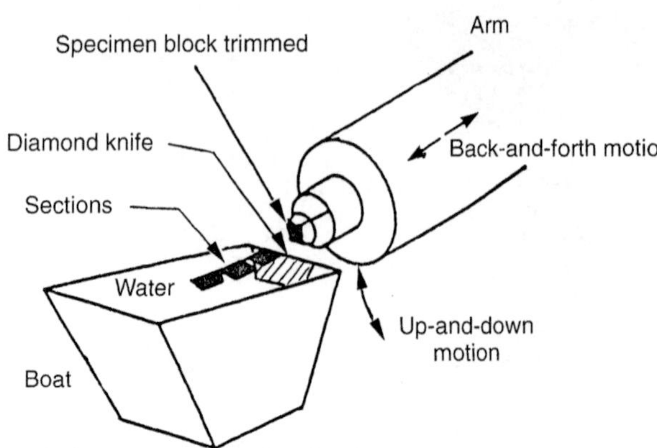

Fig. 5.42. Principle of ultramicrotomy. Hatched region corresponds to a diamond knife

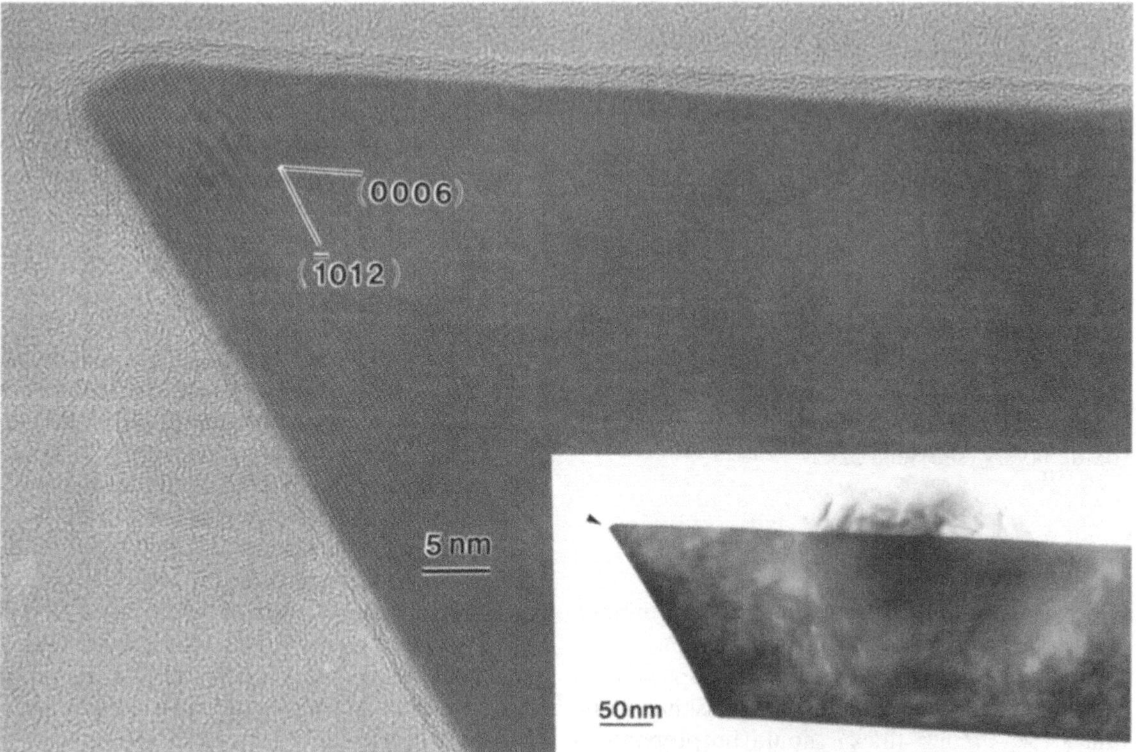

Fig. 5.43. High-resolution electron microscope image of a section of platelet-type hematite particle. The region corresponds to an area shown in the low magnified image shown at bottom right

If the trimming and the slicing are not carried out correctly, the expensive diamond knife is damaged. Also, lattice strain is frequently introduced into the sections during slicing (see the inset of Fig. 5.43).

5.5.5 Ion Milling

Ion milling is widely used to obtain thin regions, especially of ceramics, semiconductors, and multilayer films. In principle, the so-called sputtering phenomenon, where atoms are ejected from the surface by irradiating them with accelerated ions, are used. First, a thin plate (<0.1 mm) is prepared from a bulk specimen using a diamond cutter and mechanical thinning. Then a disk 3 mm in diameter is made from the plate using a diamond cutter or a ultrasonic cutter, and a dimple is formed around the center of the surface with a dimple grinder (Fig. 5.44a). If it is possible to thin the disk directly to 0.03 mm by mechanical thinning without using a dimple grinder, the disk should be strengthened by covering the edge with a metal ring such as an Mo ring (Fig. 5.44b). Ar ions are usually used for the sputtering, and the incidence angle against the disk specimen and the accelerating voltage are set as 10°–20° and several kilovolts, respectively. In a conventional ion milling system, ion milling is automatically stopped when a hole is made in the specimen by detecting the laser beam (Fig. 5.44c). When ion milling is continued for some time, the composition sometimes changes at the surface owing to the difference of sputtering efficiencies in the constituent atoms, and amorphous layers form on the surface owing to ion irradiation damage. To avoid these effects, the condition of the ion milling should be optimized (i.e., by using different ions, lowering the accelerating voltage, and adjusting the incident beam angle). To minimize any increase in specimen temperature during sputtering, use of a cooling stage with liquid nitrogen is effective. If the incidence angle becomes too small, a metal ring used for strengthening the disk is irradiated with the ions, and the specimen plate is coated with the metal. The existence of such impurity phases can be detected directly from energy-loss spectra and energy dispersive X-ray spectra. Ion milling is also used as the final thinning process to remove any contamination of a thin specimen prepared by electropolishing or chemical polishing.

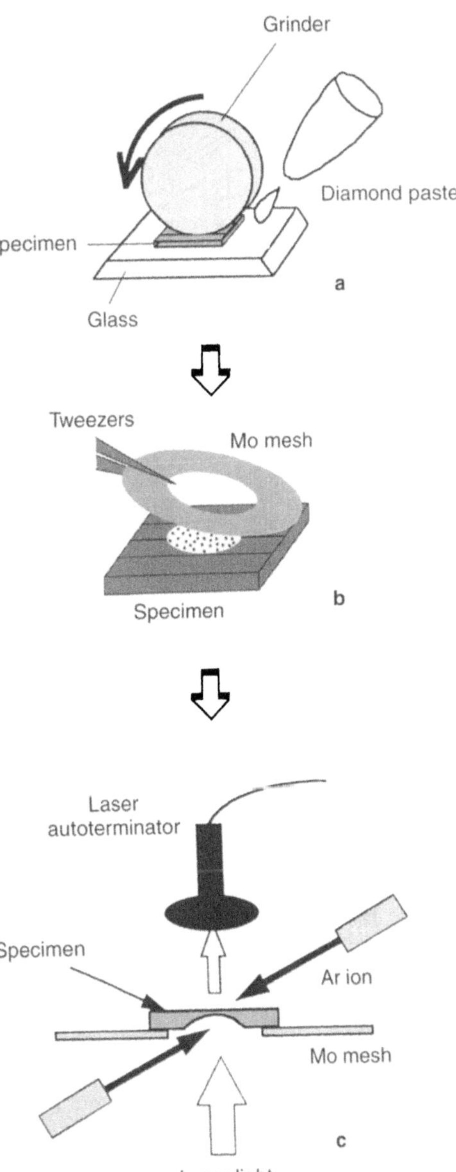

Fig. 5.44. Specimen preparation procedure for the ion milling method

5.5.6 Focused Ion Beam

The focused ion beam (FIB) method was originally developed for the purpose of fixing semiconductor devices. Ion beams are sharply focused on a small area, and the specimen is thinned rapidly by sputtering. Usually Ga ions are used with an accelerating voltage of about 30 kV and a current of about 10 A/cm^2. The probe size is several tens of nanometers. Figure 5.45a illustrates the incident beam directions of these ions and of

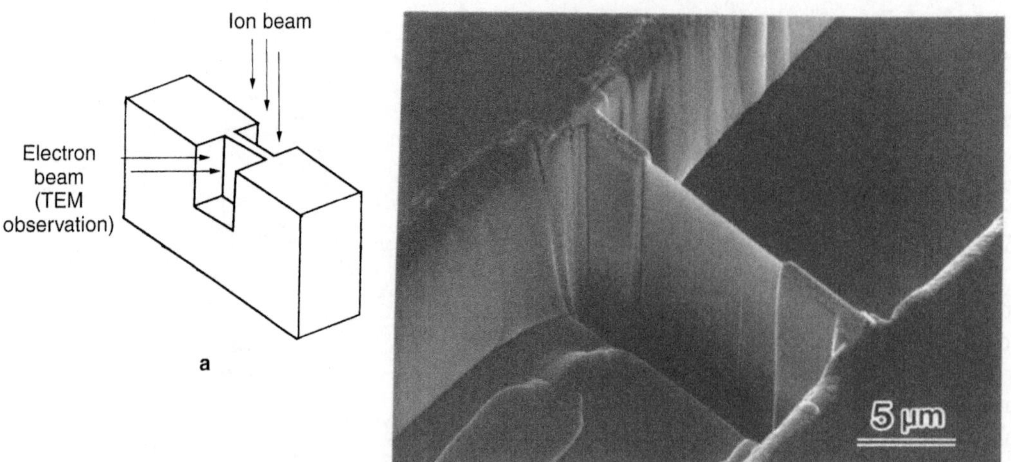

Fig. 5.45. **a** Geometric relation between the direction of an ion beam for the focused ion beam (FIB) method and the direction of electron beam for TEM observation. **b** Thin part of silicon prepared with the FIB method and observed with secondary electrons

the electrons used in the observation. Figure 5.45b shows a thin sample of silicon prepared with the FIB method [35].

This method is currently attracting much attention. It may be especially useful for specimens that contain a boundary between different materials, where it may be difficult to thin the boundary region homogeneously by other methods, such as ion milling. By detecting the secondary electrons emitted from the specimen while irradiating it with ion beams, a secondary electron image of the surface can be displayed as an SEM image. Thus, by observing the secondary electron image, one can accurately select the appropriate region for thinning. Special care should be taken to avoid irradiation damage due to the strong ion beams and to avoid implantation of Ga ions. EDS analysis is important for detecting implanted ions. The FIB system is expensive compared with other thinning instruments.

5.5.7 Vacuum Evaporation

The vacuum evaporation method is used to prepare homogeneous thin films of metals and alloys; it is also used to coat specimens with the metal or alloy. The specimen is set in a W-coil or basket. Resistance heating is applied by an electrical current passing through the coil or basket; the specimen is melted, then evaporated (or sublimated), and finally deposited onto a substrate. The deposition process is usually carried under a pressure of 10^{-3}–10^{-4} Pa, but to avoid surface contamination a higher vacuum is necessary. A collodion film or cleaved rock salt is used as a substrate. Rock salt is especially useful for forming single crystals with a special orientation relation between each crystal and the substrate. Salt is easily dissolved in water, and the deposited films can be fixed on a grid. A quartz crystal film thickness monitor can be used for accurate thickness measurements. The method is useful for preparing a standard specimen for thickness measurement.

5.5.8 Observation of Specimens

5.5.8.1 Avoiding Contamination

When a small precipitate or an interface is analyzed, a small region of nanometer scale should be illuminated with a nano-probe. In this case, attention should be paid to the contamination around the illuminated region. It is necessary to carry out such analysis under a high vacuum condition and by an anticontamination trap with liquid nitrogen. It is also effective to illuminate a wide area, including the part to be analyzed, with a strong electron beam with a large condenser aperture and a large spot size before making use of a nano-probe.

5.5.8.2 Detecting Contamination and Impurities

The occurrence of surface contamination and existence of an impurity phase can be detected by the appearance of a C-K line in EDS or the appearance of a C K-edge and the change of a low-

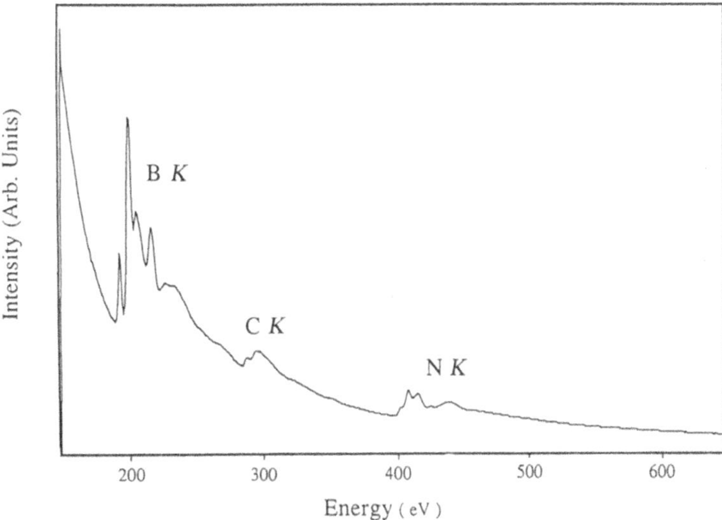

Fig. 5.46. Energy-loss spectrum of boron nitride (BN) prepared by the crushing method

loss peak in EELS. Figure 5.46 shows a part of an energy-loss spectrum of boron nitride (BN) prepared with a crushing method. There is a CK-edge due to contamination in addition to the K-edges of constituent elements B and N. Surface oxide layers can be detected by an oxygen K-edge in EELS (see Fig. 3.10); and in electron diffraction patterns (e.g., oxide layers) they can be confirmed by Debye rings from a polycrystalline phase or a halo ring from an amorphous phase.

References

1. Hirsch PB, Howie A, Nicholson RB, Pashley DW, Whelan MJ (1965) Electron microscopy of thin crystals. Butterworths, London
2. Cowley JM (1984) Diffraction physics, 2nd edn. Elsevier Science, Amsterdam
3. Shindo D, Hiraga K (1998) High-resolution electron microscopy for materials science. Springer-Verlag, Tokyo
4. Buxton BF, Eades JA, Steeds JW, Rackham GM (1976) The symmetry of electron diffraction zone axis patterns. Philos Trans R Soc Lond 281:171
5. Tanaka M, Terauchi M (1985) Convergent-beam electron diffraction. JEOL-Maruzen, Tokyo p 192
6. Yonenaga I, Brown PD, Burgess WG, Humphreys CJ (1995) Faulted dipoles in indium-doped GaAs. Inst Phys Conf Ser 146:87
7. Kelly PM, Jostsons A, Blake RG, Napier JG (1975) The determination of foil thickness by scanning transmission electron microscopy. Phys Stat Sol (a) 31:771
8. Allen SM (1981) Foil thickness measurement from convergent-beam diffraction patterns. Phil Mag A43:325
9. Nishino D, Nakafuji A, Yang J-M, Shindo D (1998) Precise morphology analysis on platelet-type hematite particles by transmission electron microscopy. ISIJ Int 38:1369
10. Cherns D, Preston AR (1986) Convergent beam diffraction studies of crystal defects. Proc 11th Int Cong Electron Microsc Kyoto 1:721
11. Gjønnes J, Høier R (1971) The application of non-systematic many-beam dynamic effects to structure-factor determination. Acta Cryst A27:313
12. Tomokiyo Y, Matsumura S, Eguchi T (1986) Critical voltage and anharmonicity of thermal vibration of atoms in metals of cubic lattices. Proc 11th Int Cong Electron Microsc Kyoto 2:1085
13. Grundy PJ, Tebble RS (1968) Lorentz electron microscopy. Adv Phys 17:153
14. Néel L (1955) Magnétisme: énergie des parois de Bloch dans les couches minces. C R Acad Sci 241:533
15. Huber EE Jr, Smith DO, Goodenough JB (1958) Domain-wall structure in permalloy films. J Appl Phys 29:294
16. Tonomura A (1999) Electron holography, 2nd edn Springer, Berlin Heidelberg New York Tokyo
17. Yang J-M, Shindo D, Lim S-H, Takeguchi M, Oikawa T (1998) Advanced transmission electron microscopy on Sm-Co based permanent magnets. Electron Microsc ICEM 14:559
18. Kittel C (1949) Physical theory of ferromagnetic domains. Rev Mod Phys 21:541
19. Fuller HW, Hale ME (1960) Determination of magnetization distribution in thin films using electron microscopy. J Appl Phys 31:238

20. Wade RH (1966) Investigation of the geometrical-optical theory of magnetic structure imaging in the electron microscope. J Appl Phys 37:366
21. Yang J-M, Shindo D, Hiroyoshi H (1997) Observation of microstructures and magnetic domains of Sm-Co based permanent magnets by high-voltage electron microscopy. Mater Transact JIM 38:363
22. Mishra RK, Thomas G, Yoneyama T, Fukuno A, Ojima T (1981) Microstructure and properties of step aged rare earth alloy magnets. J Appl Phys 52:2517
23. Chapman JN, Johnston AB, Heyderman LJ, McVitie S, Nicholson WAP, Bormans B (1994) Coherent magnetic imaging by TEM. IEEE Trans Magn 30:4479
24. McVitie S, Chapman JN, Zhou L, Heyderman LJ, Nicholson WAP (1995) In-situ magnetizing experiments using coherent magnetic imaging in TEM. J Magn Magn Mater 148:232
25. Chapman JN, Batson PE, Waddell EM, Ferrier RP (1978) The direct determination of magnetic domain wall profiles by differential phase contrast electron microscopy. Ultramicroscopy 3:203
26. Tsuno K, Inoue M (1984) Double gap objective lens for observing magnetic domains by means of differential phase contrast electron microscopy. Optik 67:363
27. Aharonov Y, Bohm D (1959) Significance of electromagnetic potentials in the quantum theory. Phys Rev 115:485
28. Shindo D, Park Y-G (2003, in press) Lorentz microscopy and holography characterization of magnetic materials. In: Characterization and simulation of advanced magnetic materials. Springer, Berlin Heidelberg New York Tokyo
29. Lee C-W, Ikematsu Y, Shindo D (2000) Thickness measurement of amorphous SiO_2 by EELS and electron holography. Mater Transact JIM 41:1129
30. Shindo D, Murakami Y, Hirayama T (1998) Application of electron hologram to morphological analysis of spindle-type hematite particles. Mater Transact JIM 39:322
31. Shindo D, Park Y-G, Yoshizawa Y (2002) Magnetic domain structures of $Fe_{73.5}Cu_1Nd_3Si_{13.5}B_9$ films studied by electron holography. J Magn Magn Mater 238:101
32. Pennycook SJ, Berger SD, Culbertson RJ (1986) Elemental mapping with elastically scattered electrons. J Microsc 144:229
33. Kawasaki M, Yamazaki T, Sato S, Watanabe K, Shiojiri M (2001) Atomic-scale quantitative elemental analysis of boundary layers in a $SrTiO_3$ ceramic condenser by high-angle annular dark-field electron microscopy. Phil Mag A 81:245
34. Shindo D, Lee B-T, Waseda Y, Muramatsu A, Sugimoto T (1993) Crystallography of platelet-type hematite particles by electron microscopy. Mater Transact JIM 34:580
35. Ikematsu Y, Mizutani T, Nakai K, Fujinami M, Hasebe M, Ohashi W (1998) Transmission electron microscope observation of grown-in defects detected by bright-field infrared-laser interferometer in Czochralshi silicon crystals. Jpn J Appl Phys 37:L196

Appendix A: Physical Constants, Conversion Factors, Electron Wavelengths

Table A.1. Physical constants.

Parameter	Constant	SI	CGS
Elementary electric charge (e)	1.6022	$\times 10^{-19}$ C	$\times 10^{-20}$ emu
	4.8032		$\times 10^{-10}$ esu
Electron mass (m_0)	9.1094	$\times 10^{-31}$ kg	$\times 10^{-28}$ g
Proton mass (m_p)	1.6726	$\times 10^{-27}$ kg	$\times 10^{-24}$ g
Neutron mass (m_n)	1.6749	$\times 10^{-27}$ kg	$\times 10^{-24}$ g
Velocity of light (c)	2.9979	$\times 10^{8}$ m·s^{-1}	$\times 10^{10}$ cm·s^{-1}
Mass energy of an electron (m_0c^2)	8.1871	$\times 10^{-14}$ J	$\times 10^{-7}$ erg
	(0.51100 MeV)		
Plank's constant (h)	6.6261	$\times 10^{-34}$ J·s	$\times 10^{-27}$ erg·s
$\hbar = h/2\pi$	1.0546	$\times 10^{-34}$ J·s	$\times 10^{-27}$ erg·s
Bohr radius (A_0)	5.2918	$\times 10^{-11}$ m	$\times 10^{-9}$ cm
Compton wavelength ($\lambda_c = h/m_0c$)	2.4263	$\times 10^{-12}$ m	$\times 10^{-10}$ cm
Avogadro's number (N_0)	6.0221	$\times 10^{23}$ mol^{-1}	$\times 10^{23}$ mol^{-1}

Table A.2. Conversion factors.

1 eV = 1.6022×10^{-19} J	1 Å = 0.1 nm	1 G = 10^{-4} T
1 torr = 133.32 Pa	1 kX = 0.10020 nm	1 Oe = $\frac{10^3}{4\pi}$ Am^{-1} = 79.577 Am^{-1}

Table A.3. Electron wavelengths and relativistic correction factors.

Accelerating voltage (kV)	Wavelength λ (nm)	$\beta_r = v/c$	$\beta_m = (1 - \beta_r^2)^{-1/2}$
80	0.00417572	0.50240	1.1566
100	0.00370144	0.54822	1.1957
120	0.00334922	0.58667	1.2348
150	0.00295704	0.63432	1.2935
180	0.00266550	0.67315	1.3523
200	0.00250793	0.69531	1.3914
300	0.00196875	0.77653	1.5871
400	0.00164394	0.82787	1.7828
500	0.00142126	0.86286	1.9785
600	0.00125680	0.88795	2.1742
700	0.00112928	0.90661	2.3699
800	0.00102695	0.92091	2.5656
900	0.00094269	0.93212	2.7613
1000	0.00087192	0.94108	2.9570
1250	0.00073571	0.95697	3.4462
1300	0.00071361	0.95937	3.5440
1500	0.00063745	0.96718	3.9354
2000	0.00050432	0.97907	4.9139
2500	0.00041783	0.98549	5.8924
3000	0.00035693	0.98935	6.8709

Appendix B: Electron Binding Energies and Characteristic X-ray Energies

Table B.1. Electron binding energies (unit: eV).

	$1s_{1/2}$ K	$2s_{1/2}$ L_{I}	$2p_{1/2}$ L_{II}	$2p_{3/2}$ L_{III}	$3s_{1/2}$ M_{I}	$3p_{1/2}$ M_{II}	$3p_{3/2}$ M_{III}	$3d_{3/2}$ M_{IV}	$3d_{5/2}$ M_{V}	$4s_{1/2}$ N_{I}	$4p_{1/2}$ N_{II}	$4p_{3/2}$ N_{III}	$4d_{3/2}$ N_{IV}	$4d_{5/2}$ N_{V}	$4f_{5/2}$ N_{VI}	$4f_{7/2}$ N_{VII}
1 H	14															
2 He	25															
3 Li	55															
4 Be	111															
5 B	188[1]		5													
6 C	284[1]		7													
7 N	399[19]		9													
8 O	532[1]	24[2]	7													
9 F	686[1]	31	9													
10 Ne	867	45	18													
11 Na	1 072[1]	63[3]	31		1											
12 Mg	1 305[1]	89[3]	52		2											
13 Al	1 560[1]	118[3]	74	73	1											
14 Si	1 839[1]	149[3]	100	99	8	3										
15 P	2 149[1]	189[3]	136	135	16	10										
16 S	2 472[1]	229[3]	165	164	16	8										
17 Cl	2 823[1]	270[3]	202	200	18	7										
18 A	3 203	320[3]	247	245	25	12										
19 K	3 608[1]	377[3]	297	294	34	18										
20 Ca	4 038[1]	438[3]	350	347	44	26		5								
21 Sc	4 493[1]	500[3]	407	402	54	32		7								
22 Ti	4 965[1]	564[3]	461	455	59	34		3								
23 V	5 465[1]	628[3]	520	513	66	38		2								
24 Cr	5 989[1]	695[3]	584	575	74	43		2								
25 Mn	6 539[4]	769[3]	652	641	84	49		4								
26 Fe	7 114[5]	846[3]	723	710	95	56		6								
27 Co	7 709[4]	926[3]	794	779	101	60		3								
28 Ni	8 333[4]	1 008[3]	872	855	112	68		4								
29 Cu	8 979[6]	1 096[3]	951	931	120	74		2								
30 Zn	9 659[6]	1 194	1 044	1 021	137	87		9								
31 Ga	10 367[6]	1 298	1 143	1 116	158	107	103	18			1					
32 Ge	11 104[6]	1 413[6]	1 249	1 217	181	129	122	29			3					
33 As	11 867[6]	1 527	1 359	1 323	204	147	141	41			3					
34 Se	12 658[6]	1 654	1 476	1 436	232	168	162	57			6					
35 Br	13 474	1 782[7]	1 596[7]	1 550[7]	257	189	182	70	69	27	5					
36 Kr	14 326	1 921	1 727	1 675	289*	223	214	89		24	11					
37 Rb	15 200	2 065[7]	1 864[7]	1 805[7]	322	248	239	112	111	30	15	14				
38 Sr	16 105	2 216[8]	2 007[8]	1 940[8]	358	280	269	135	133	38	20					
39 Y	17 039	2 373[8]	2 155[8]	2 080[8]	395	313	301	160	158	46	26		3			
40 Zr	17 998	2 532[8]	2 307[8]	2 223[8]	431	345	331	183	180	52	29		3			
41 Nb	18 986	2 698[8]	2 465[8]	2 371[8]	469	379	363	208	205	58	34		4			
42 Mo	20 000	2 866[8]	2 625[8]	2 520[8]	505	410	393	230	227	62	35		2			
43 Tc	21 044	3 043[8]	2 793[8]	2 677[8]	544*	445	425	257	253	68*	39		2*			
44 Ru	22 117	3 224[8]	2 967[8]	2 838[8]	585	483	461	284	279	75	43		2			
45 Rh	23 220	3 412[8]	3 146[8]	3 004[8]	627	521	496	312	307	81	48		3			

* Energies obtained by interpolation

† Extrapolated energies or energies obtained by self-consistent field calculation

This table is from ESCA (Almqvist and Wiksells Boktryckeri AB, 1967), through the courtesy of Prof. K. Siegbahn

	$1s_{1/2}$ K	$2s_{1/2}$ L_{I}	$2p_{1/2}$ L_{II}	$2p_{3/2}$ L_{III}	$3s_{1/2}$ M_{I}	$3p_{1/2}$ M_{II}	$3p_{3/2}$ M_{III}	$3d_{3/2}$ M_{IV}	$3d_{5/2}$ M_{V}	$4s_{1/2}$ N_{I}	$4p_{1/2}$ N_{II}	$4p_{3/2}$ N_{III}	$4d_{3/2}$ N_{IV}	$4d_{5/2}$ N_{V}	$4f_{5/2}$ N_{VI}	$4f_{7/2}$ N_{VII}
46 Pd	24 350	3 605[8]	3 331[8]	3 173[8]	670	559	531	340	335	86	51		1			
47 Ag	25 514	3 806[9]	3 524[6,9]	3 351[6]	717	602	571	373	367	95	62	56	3			
48 Cd	26 711	4 018[9]	3 727[9]	3 538[9]	770	651	617	411	404	108	67		9			
49 In	27 940	4 238[9]	3 938[9]	3 730[6]	826	702	664	451	443	122	77		16			
50 Sn	29 200	4 465[6]	4 156[6]	3 929[6]	884	757	715	494	485	137	89		24			
51 Sb	30 491	4 699[9]	4 381[9]	4 132[6]	944	812	766	537	528	152	99		32			
52 Te	31 814	4 939[9]	4 612[9]	4 341[6]	1 006	870	819	582	572	168	110		40			
53 I	33 170	5 188[7]	4 852[7]	4 557[7]	1 072	931	875	631	620	186	123		50			
54 Xe	34 561	5 453	5 104	4 782	1 145*	999	937	685*	672	208*	147		63*			
55 Cs	35 985	5 713[7]	5 360[7]	5 012[7]	1 217	1 065	998	740	726	231	172	162	79	77		
56 Ba	37 441	5 987[7]	5 624[7]	5 247[7]	1 293	1 137	1 063	796	781	253	192	180	93	90		
57 La	38 925	6 267[10]	5 891[10]	5 483[10]	1 362	1 205	1 124	849	832	271	206	192	99			
58 Ce	40 444	6 549[10]	6 165[10]	5 724[10]	1 435	1 273	1 186	902	884	290	224	208	111		1	
59 Pr	41 991	6 835[10]	6 441[10]	5 965[10]	1 511	1 338	1 243	951	931	305	237	218	114		2	
60 Nd	43 569	7 126[10]	6 722[10]	6 208[10]	1 576	1 403	1 298	1 000	978	316	244	225	118		2	
61 Pm	45 185*	7 428*[11]	7 013*[11]	6 460*[11]	1 650*	1 472*	1 357*	1 052*	1 027*	331*	255*	237*	121*		4*	
62 Sm	46 835	7 737[10]	7 312[10]	6 717[10]	1 724	1 542	1 421	1 107	1 081	347	267	249	130		7	
63 Eu	48 519	8 052[10]	7 618[10]	6 977[10]	1 800	1 614	1 481	1 161	1 131	360	284	257	134		0	
64 Gd	50 239	8 376[10]	7 931[10]	7 243[10]	1 881	1 689	1 544	1 218	1 186	376	289	271	141		0	
65 Tb	51 996	8 708[10]	8 252[10]	7 515[10]	1 968	1 768	1 612	1 276	1 242	398	311	286	148		3	
66 Dy	53 788	9 047[10]	8 581[10]	7 790[10]	2 047	1 842	1 676	1 332	1 295	416	332	293	154		4	
67 Ho	55 618	9 395[10]	8 919[10]	8 071[10]	2 128	1 923	1 741	1 391	1 351	436	343	306	161		4	
68 Er	57 486	9 752[10]	9 265[10]	8 358[10]	2 207	2 006	1 812	1 453	1 409	449	366	320	177	168	4	
69 Tm	59 390	10 116[10]	9 618[10]	8 648[10]	2 307	2 090	1 885	1 515	1 468	472	386	337	180		5	
70 Yb	61 332	10 488[10]	9 978[10]	8 943[10]	2 397	2 172	1 949	1 576	1 527	487	396	343	197	184	6	
71 Lu	63 314	10 870[10]	10 349[10]	9 244[10]	2 491	2 264	2 024	1 640	1 589	506	410	359	205	195	7	
72 Hf	65 351	11 272[11]	10 739[11]	9 561[11]	2 601	2 365	2 108	1 716	1 662	538	437	380	224	214	19[17]	18[17]
73 Ta	67 417	11 680[12]	11 136[12]	9 881[12]	2 708	2 469[12]	2 194[12]	1 793[12]	1 735[12]	566	465	405	242	230	27[17]	25[17]
74 W	69 525	12 099[12]	11 542[12]	10 205[12]	2 820	2 575[12]	2 281[12]	1 872[12]	1 810[12]	595	492	426	259	246	37[17]	34[17]
75 Re	71 677	12 527	11 957[12]	10 535[12]	2 932	2 682[12]	2 367[12]	1 949[12]	1 883[12]	625	518	445	274	260	47[17]	45[17]
76 Os	73 871	12 968	12 385	10 871[12]	3 049	2 792[12]	2 458[12]	2 031[12]	1 960[12]	655	547	469	290	273	52[17]	50[17]
77 Ir	76 111	13 419	12 824[12]	11 215[12]	3 174	2 909[12]	2 551[12]	2 116[12]	2 041[12]	690	577	495	312	295	63[17]	60[17]
78 Pt	78 395	13 880[20]	13 273[20]	11 564[20]	3 298[20]	3 027[20]	2 646[20]	2 202[20]	2 121[20]	724[20]	608[20]	519[20]	331[20]	314[20]	74[20]	70[20]
79 Au	80 725	14 353	13 733[12]	11 918[12]	3 425[12]	3 150[12]	2 743[12]	2 291[12]	2 206[12]	759	644	546	352	334	87[17]	83[17]
80 Hg	83 103	14 839	14 209	12 284[12]	3 562	3 279	2 847[12]	2 385[12]	2 295[12]	800	677	571	379	360	103[17]	99[17]
81 Tl	85 531	15 347	14 698[12]	12 657[12]	3 704	3 416[12]	2 957[12]	2 485[12]	2 390[12]	846	722	609	407	386	122[17]	118[17]
82 Pb	88 005	15 861	15 200	13 035[12]	3 851	3 554[12]	3 067[12]	2 586[12]	2 484[12]	894	764	645	435	413	143[17]	138[17]
83 Bi	90 526	16 388	15 709[12]	13 418[12]	3 999[12]	3 697[12]	3 177[12]	2 688[12]	2 580[12]	939[12]	806[12]	679[12]	464[12]	440[12]	163[17]	158[17]
84 Po	93 105	16 939	16 244	13 814	4 149	3 854	3 302	2 798	2 683	995	851	705	500	473	184*	
85 At	95 730	17 493	16 785	14 214	4 317*	4 008	3 426	2 909	2 787	1 042*	886	740	533	507*	210*	
86 Rn	98 404	18 049	17 337	14 619	4 482*	4 159	3 538	3 022	2 892	1 097*	929	768	567	541*	238*	
87 Fr	101 137	18 639	17 906	15 031	4 652*	4 327*	3 663	3 136	3 000	1 153*	980	810	603	577	268*	
88 Ra	103 922	19 237	18 484	15 444	4 822*	4 490	3 792	3 248	3 105	1 208*	1 058	879	636	603	299	
89 Ac	106 755	19 840	19 083	15 871	5 002	4 656	3 909	3 370	3 219	1 269*	1 080	890	675	639*	319*	
90 Th	109 651	20 472	19 693	16 300[13]	5 182[13]	4 831[13]	4 046[13]	3 491[13]	3 332[13]	1330[13]	1168[13]	968[13]	714[13]	677[13]	344[13]	335[13]
91 Pa	112 601	21 105	20 314	16 733	5 367	5 001	4 174	3 611	3 442	1 387	1 224	1 007	743	708	371	360
92 U	115 606	21 758	20 948	17 168[14]	5 548	5 181[14]	4 304[14]	3 728[14]	3 552[14]	1442[14]	1273[14]	1045[14]	780[14]	738[14]	392	381
93 Np	118 676	22 420	21 599	17 608	5 722	5 366[15]	4 435[15]	3 850[15]	3 664[15]	1501[15]	1328[15]	1087[15]	817[15]	773[15]	415[15]	404[15]
94 Pu	121 818	23 102	22 266	18 057	5 933	5 546	4 562	3 973[16]	3 778[16]	1 558	1 377	1 120	849[16]	801[16]	422	
95 Am	125 027	23 773	22 944	18 504	6 120	5 710	4 667	4 092	3 887	1 617	1 412	1 136+	879	828	440+	
96 Cm	128 220*	24 460*	23 779*	18 930*	6 288*	5 895*	4 797*	4 227+	3 971+	1 643*	1 440+	1 154+				
97 Bk	131 590[18]	25 275[18]	24 385[18]	19 452[18]	6 556[18]	6 147[18]	4 977[18]	4 366+	4 132+	1755[18]	1 554+	1 235+				
98 Cf	135 960+	26 110+	25 250+	19 930+	6 754+	6 359+	5 109+	4 497+	4 253+	1 791+	1 616+	1 279+				
99 Es	139 490+	26 900+	26 020+	20 410+	6 977+	6 574+	5 252+	4 630+	4 374+	1 868+	1 680+	1 321+				
100 Fm	143 090+	27 700+	26 810+	20 900+	7 205+	6 793+	5 397+	4 766+	4 498+	1 937+	1 747+	1 366+				
101 Md	146 780+	28 530+	27 610+	21 390+	7 441+	7 019+	5 546+	4 903+	4 622+	2 010+	1 814+	1 410+				
102 No	150 540+	29 380+	28 440+	21 880+	7 675+	7 245+	5 688+	5 037+	4 741+	2 078+	1 876+	1 448+				
103 Lr	154 380+	30 240+	29 280+	22 360+	7 900+	7 460+	5 810+	5 150+	4 860+	2 140+	1 930+	1 480+				
104 Ku	158 300+	31 120+	30 140+	22 840+	8 120+	7 660+	5 910+	5 240+	4 980+	2 200+	1 970+	1 510+				

$5s_{1/2}$ O_{I}	$5p_{1/2}$ O_{II}	$5p_{3/2}$ O_{III}	$5d_{3/2}$ O_{IV}	$5d_{5/2}$ O_{V}	$6s_{1/2}$ P_{I}	$6p_{1/2}$ P_{II}	$6p_{3/2}$ P_{III}	$6d_{3/2}$ P_{IV}	$6d_{5/2}$ P_{V}
	2								
	1								
1	1								
7	2								
12	2								
14	3								
18*	7*								
23	13	12							
40	17	15							
33	15								
38	20								
38	23								
38	22								
38*	22*								
39	22								
32	22								
36	21								
40	26								
63	26								
51	20								
60	29								
53	32								
53	23								
57	28		5						
65	38	31	7						
71	45	37	6						
77	47	37	6						
83	46	35	4						
84	58	46	0						
96	63	51	4						
102	66	51[20]	2[20]						
108	72	54	3						
120	81	58	7						
137	100	76	16	13					
148	105	86	22	20	3	1			
160	117	93	27	25	8*	3			
177*	132*	104*	31		12*	5*			
195*	148*	115*	40*		18*	8*			
214*	164*	127*	48*		26*	11*			
234*	182*	140*	58*		34*	15*			
254	200	153	68		44	19			
272*	215*	167*	80*						
290	229	182[13]	95[13]	88[13]	60	49	43	2	
310	223		94						
324	260	195	105	96	71	43	33	4	
338*	283[15]	206[15]	109[15]	101[15]					
352	279	212	116	105					
367*	290+	220+	116	103					
382*									
398[18]									
419+									
435+									
454+									
472+									
484+									
490+									
500+									

Table B.2. Characteristic X-ray energies (unit: keV).

	K-SERIES						L-SERIES: L_I-SERIES				L-SERIES: L_{II}-SERIES				L-SERIES: L_{III}-SERIES					M-SERIES: M_{IV}-SERIES		M-SERIES: M_V-SERIES			
Z Element	K(ab)	Kβ3	Kβ1	Kβ2	Kα1	Kα2	L_I(ab)	Lγ3	Lβ3	Lβ4	L_{II}(ab)	Lγ1	Lβ1	Lη	L_{III}(ab)	Lβ2	Lα1	Lα2	Ll	M_{IV}(ab)	Mβ	M_V(ab)	Mα1	Mα2	Z Element
1 H	0.0136																								1 H
2 He	0.025																								2 He
3 Li	0.055				0.052																				3 Li
4 Be	0.112				0.110																				4 Be
5 B	0.192				0.185																				5 B
6 C	0.283				0.277																				6 C
7 N	0.399				0.392																				7 N
8 O	0.531				0.525																				8 O
9 F	0.687				0.677																				9 F
10 Ne	0.867				0.848																				10 Ne
11 Na	1.072		1.067		1.041																				11 Na
12 Mg	1.305		1.295		1,253																				12 Mg
13 Al	1.559		1.553		1.486	1.486																			13 Al
14 Si	1.838		1.829		1.740	1.739																			14 Si
15 P	2.142		2.136		2.013	2.012																			15 P
16 S	2.472		2.464		2.307	2.306																			16 S
17 Cl	2.822				2.622	2.620																			17 Cl
18 Ar	3.202	3.190			2.957	2.955																			18 Ar
19 K	3.607	3.589			3.313	3,310								0.262					0.260						19 K
20 Ca	4.038	4.012			3.691	3.687	0.400					0.350	0.345	0.306	0.346		0.341		0.303						20 Ca
21 Sc	4.496	4.460			4.090	4.085	0.463					0.407	0.400	0.353	0.403		0.395		0.348						21 Sc
22 Ti	4.965	4.931			4.510	4.504	0.530					0.460	0.458	0.401	0.454		0.452		0.395						22 Ti
23 V	5.465	5.426			4.951	4.944	0.604		0.585			0.520	0.519	0.453	0.513		0.511		0.446						23 V
24 Cr	5.989	5.946			5.414	5.405	0.682		0.654			0.583	0.583	0.510	0.574		0.573		0.500						24 Cr
25 Mn	6.540	6.489			5.898	5.887	0.754		0.721			0.652	0.649	0.567	0.641		0.637		0.556						25 Mn
26 Fe	7.112	7.057			6.403	6.390	0.842		0.792			0.721	0.718	0.628	0.709		0.705		0.615						26 Fe
27 Co	7.709	7.648			6.929	6.914	0.929		0.866			0.794	0.791	0.694	0.779		0.776		0.678						27 Co
28 Ni	8.333	8.263			7.477	7.460	1.012		0.941			0.872	0.869	0.762	0.855		0.851		0.743						28 Ni
29 Cu	8.979	8.904			8.046	8.026	1,100		1.023			0.952	0.950	0.832	0.932		0.930		0.811						29 Cu
30 Zn	9.659	9.570		9.656	8.637	8.614	1.196		1,107			1.044	1.034	0.906	1.021		1.012		0.884						30 Zn

This table is from *XES* (*X-ray Energy Spectrometry*, Rolf Woldseth, published by Kevex Corporation, 1973).
K(ab), L_I(ab), and so on indicate the binding energies of the electrons in the K shell, L_I shell and so on.

	K-SERIES						L-SERIES														M-SERIES					
							L_I-SERIES				L_{II}-SERIES				L_{III}-SERIES					M_{IV}-SERIES		M_V-SERIES			
Z Element	K(ab)	Kβ3	Kβ1	Kβ2	Kα1	Kα2	L_I(ab)	Lγ3	Lβ3	Lβ4	L_{II}(ab)	Lγ1	Lβ1	Lη	L_{III}(ab)	Lβ2	Lα1	Lα2	Ll	M_{IV}(ab)	Mβ	M_V(ab)	Mα1	Mα2	Z Element
31 Ga	10.368	10.259	10.263	10.365	9.250	9.223	1.300		1.197			1.134	1.125	0.984	1.117		1.098		0.957						31 Ga
32 Ge	11.104	10.976	10.980	11.099	9.885	9.854	1.420		1.294	1.286		1.249	1.218	1.068	1.218		1.188		1.036						32 Ge
33 As	11.868	11.718	11.724	11.862	10.542	10.506	1.530		1.388			1.360	1.317	1.155	1.325		1.282		1.120						33 As
34 Se	12.658	12.437	12.494	12.650	11.220	11.179	1.653		1.490			1.477	1.419	1.244	1.436		1.379		1.204						34 Se
35 Br	13.474	13.282	13.289	13.467	11.922	11.876	1.794		1.596		1.596		1.526	1.339	1.550		1.480		1.293						35 Br
36 Kr	14.322	14.102	14.110	14.312	12.648	12.596	1.920		1.706	1.697	1.756		1.636		1.675		1.586								36 Kr
37 Rb	15.201	14.949	14.959	15.183	13.393	13.333	2.067		1.826	1.817	1.866		1.752	1.542	1.806		1.694	1.692	1.482						37 Rb
38 Sr	16.105	15.822	15.833	16.082	14.163	14.095	2.216		1.947	1.936	2.007		1.871	1.649	1.940		1.806	1.804	1.582						38 Sr
39 Y	17.037	16.723	16.735	17.013	14.956	14.880	2.369		2.072	2.060	2.145		1.995	1.761	2.079		1.922	1.920	1.685						39 Y
40 Zr	17.998	17.651	17.665	17.967	15.772	15.688	2.547		2.201	2.187	2.307	2.302	2.124	1.876	2.223	2.219	2.042	2.040	1.792						40 Zr
41 Nb	18.986	18.603	18.619	18.949	16.612	16.518	2.698		2.334	2.319	2.465	2.461	2.257	1.996	2.371	2.367	2.166	2.163	1.902						41 Nb
42 Mo	20.002	19.587	19.605	19.962	17.476	17.371	2.866		2.473	2.455	2.625	2.623	2.394	2.120	2.520	2.518	2.293	2.289	2.015						42 Mo
43 Tc	21.054	20.595	20.615	21.002	18.364	18.248	3.054				2.795		2.536		2.677		2.424								43 Tc
44 Ru	22.118	21.631	21.653	22.070	19.276	19.147	3.236		2.763	2.741	2.966	2.964	2.683	2.382	2.837	2.835	2.558	2.554	2.252						44 Ru
45 Rh	23.224	22.695	22.720	23.169	20.213	20.070	3.419		2.915	2.890	3.146	3.143	2.834	2.519	3.003	3.001	2.696	2.692	2.376						45 Rh
46 Pd	24.350	23.787	23.815	24.295	21.174	21.017	3.617		3.072	3.045	3.330	3.328	2.990	2.660	3.173	3.171	2.838	2.833	2.503						46 Pd
47 Ag	25.514	24.907	24.938	25.452	22.159	21.987	3.806	3.749	3.234	3.203	3.524	3.519	3.150	2.806	3.351	3.347	2.984	2.978	2.633						47 Ag
48 Cd	26.711	26.057	26.091	26.639	23.170	22.980	4.019		3.401	3.367	3.727	3.716	3.316	2.956	3.537	3.528	3.133	3.126	2.767						48 Cd
49 In	27.940	27.233	27.271	27.856	24.206	23.998	4.237		3.572	3.535	3.938	3.920	3.487	3.112	3.730	3.713	3.286	3.279	2.904						49 In
50 Sn	29.200	28.439	28.481	29.104	25.267	25.040	4.465		3.750	3.708	4.156	4.130	3.662	3.272	3.929	3.904	3.443	3.435	3.044						50 Sn
51 Sb	30.491	29.674	29.721	30.388	26.355	26.106	4.698		3.932	3.886	4.381	4.347	3.843	3.436	4.132	4.100	3.604	3.595	3.188						51 Sb
52 Te	31.813	30.939	30.990	31.698	27.468	27.197	4.939		4.120	4.069	4.612	4.570	4.029	3.605	4.341	4.301	3.769	3.758	3.335						52 Te
53 I	33.169	32.234	32.289	33.036	28.607	28.312	5.188		4.313	4.257	4.852	4.800	4.220	3.780	4.557	4.507	3.937	3.925	3.484						53 I
54 Xe	34.582	33.556	33.619	34.408	29.774	29.453	5.452				5.100				4.781		4.109								54 Xe
55 Cs	35.959	34.913	34.981	35.815	30.968	30.620	5.720	5.552	4.716	4.649	5.358	5.279	4.619	4.141	5.011	4.935	4.286	4.272	3.794						55 Cs
56 Ba	37.441	36.298	36.372	37.251	32.188	31.812	5.995	5.808	4.926	4.851	5.624	5.530	4.827	4.330	5.247	5.156	4.465	4.450	3.953						56 Ba
57 La	38.925	37.714	37.795	38.723	33.436	33.028	6.267	6.073	5.143	5.061	5.891	5.788	5.041	4.524	5.483	5.383	4.650	4.633	4.124	0.851	0.854				57 La
58 Ce	40.449	39.163	39.251	40.226	34.714	34.273	6.549	6.340	5.364	5.276	6.165	6.051	5.261	4.731	5.724	5.612	4.839	4.822	4.287	0.902	0.902				58 Ce
59 Pr	41.998	40.646	40.741	41.767	36.020	35.544	6.846	6.615	5.591	5.497	6.443	6.321	5.488	4.935	5.968	5.849	5.033	5.013	4.452		0.949				59 Pr
60 Nd	43.571	42.159	42.264	43.327	37.355	36.841	7.126	6.900	5.828	5.721	6.722	6.601	5.721	5.145	6.208	6.088	5.229	5.207	4.632	1.004	0.996				60 Nd
61 Pm	45.207	43.705	43.818	44.929	38.718	38.165	7.448		6.070		7.018	6.891	5.960		6.466	6.338	5.432	5.407							61 Pm
62 Sm	46.835	45.281	45.405	46.566	40.111	39.516	7.737	7.485	6.317	6.195	7.312	7.177	6.205	5.588	6.717	6.586	5.635	5.607	4.994	1.108	1.100				62 Sm
63 Eu	48.515	46.896	47.030	48.248	41.535	40.895	8.069	7.795	6.570	6.438	7.624	7.479	6.455	5.816	6.983	6.842	5.845	5.816	5.176		1.153				63 Eu
64 Gd	50.240	48.547	48.688	49.952	42.989	42.302	8.376	8.104	6.830	6.686	7.931	7.784	6.712	6.049	7.243	7.102	6.056	6.024	5.361	1.221	1.209				64 Gd

	K-SERIES						L-SERIES													M-SERIES					
							L_I-SERIES				L_{II}-SERIES				L_{III} SERIES					M_{IV}-SERIES		M_V-SERIES			
Z Element	K(ab)	$K\beta3$	$K\beta1$	$K\beta2$	$K\alpha1$	$K\alpha2$	L_I(ab)	$L\gamma3$	$L\beta3$	$L\beta4$	L_{II}(ab)	$L\gamma1$	$L\beta1$	$L\eta$	L_{III}(ab)	$L\beta2$	$L\alpha1$	$L\alpha2$	Ll	M_{IV}(ab)	$M\beta$	M_V(ab)	$M\alpha1$	$M\alpha2$	Z Element
65 Tb	51.996	50.221	50.374	51.715	44.474	43.737	8.708	8.422	7.095	6.939	8.252	8.100	6.977	6.283	7.515	7.365	6.272	6.237	5.546	1.280	1.266				65 Tb
66 Dy	53.789	51.949	52.110	53.500	45.991	45.200	9.083	8.752	7.369	7.203	8.621	8.417	7.246	6.533	7.850	7.634	6.494	6.457	5.742		1.325				66 Dy
67 Ho	55.615	53.702	53.868	55.315	47.539	46.692	9.395	9.086	7.650	7.470	8.919	8.746	7.524	6.787	8.071	7.910	6.719	6.679	5.942	1.390	1.383				67 Ho
68 Er	57.483	55.485	55.672	57.204	49.119	48.213	9.776	9.429	7.938	7.744	9.263	9.087	7.809	7.057	8.364	8.188	6.947	6.904	6.152		1.443				68 Er
69 Tm	59.390	57.293	57.506	59.085	50.733	49.764	10.116	9.778	8.229	8.024	9.618	9.424	8.100	7.308	8.648	8.467	7.179	7.132	6.341	1.515	1.503				69 Tm
70 Yb	61.332	59.141	59.356	60.974	52.380	51.345	10.486	10.141	8.535	8.312	9.978	9.778	8.400	7.579	8.943	8.757	7.414	7.366	6.544	1.578	1.567				70 Yb
71 Lu	63.304	61.037	61.272	62.956	54.061	52.956	10.867	10.509	8.845	8.605	10.345	10.142	8.708	7.856	9.241	9.047	7.654	7.604	6.752		1.631				71 Lu
72 Hf	65.351	62.969	63.222	64.969	55.781	54.602	11.264	10.889	9.162	8.904	10.739	10.514	9.021	8.138	9.561	9.346	7.898	7.843	6.958	1.718	1.697				72 Hf
73 Ta	67.414	64.938	65.212	67.001	57.523	56.267	11.680	11.276	9.486	9.211	11.139	10.893	9.342	8.427	9.881	9.650	8.145	8.086	7.172	1.793	1.765				73 Ta
74 W	69.524	66.940	67.233	69.089	59.308	57.972	12.098	11.672	9.817	9.524	11.542	11.284	9.671	8.723	10.204	9.960	8.396	8.334	7.386	1.871	1.835	1.809	1.775	1.773	74 W
75 Re	71.662	68.983	69.298	71.219	61.130	59.708	12.522	12.080	10.158	9.845	11.955	11.683	10.008	9.026	10.531	10.274	8.651	8.585	7.602		1.906				75 Re
76 Os	73.860	71.065	71.401	73.390	62.990	61.476	12.965	12.498	10.509	10.174	12.383	12.093	10.354	9.335	10.869	10.597	8.910	8.840	7.821		1.978				76 Os
77 Ir	76.112	73.190	73.548	75.606	64.885	63.276	13.424	12.922	10.866	10.509	12.824	12.510	10.706	9.649	11.215	10.919	9.174	9.098	8.040	2.116	2.053	2.041	1.980	1.975	77 Ir
78 Pt	78.395	75.355	75.735	77.864	66.821	65.112	13.892	13.359	11.233	10.852	13.273	12.940	11.069	9.973	11.564	11.249	9.441	9.360	8.267	2.202	2.127	2.122	2.050	2.046	78 Pt
79 Au	80.723	77.567	77.971	80.172	68.792	66.978	14.353	13.807	11.608	11.203	13.733	13.379	11.440	10.307	11.918	11.583	9.712	9.626	8.493	2.291	2.204	2.206	2.123	2.118	79 Au
80 Hg	63.103	79.809	80.240	82.530	70.807	68.883	14.846	14.262	11.993	11.561	14.209	13.828	11.821	10.649	12.284	11.922	9.987	9.896	8.720	2.385	2.282	2.295			80 Hg
81 Tl	85.528	82.104	82.562	84.933	72.859	70.820	15.344	14.734	12.388	11.929	14.698	14.289	12.211	10.992	12.657	12.270	10.267	10.171	8.952	2.485	2.362	2.389	2.270	2.265	81 Tl
82 Pb	88.006	84.436	84.922	87.351	74.956	72.792	15.860	15.215	12.791	12.304	15.198	14.762	12.612	11.347	13.035	12.621	10.550	10.448	9.183	2.586	2.442	2.484	2.345	2.339	82 Pb
83 Bi	90.527	86.819	87.328	89.846	77.095	74.802	16.385	15.708	13.208	12.689	15.708	15.245	13.021	11.710	13.418	12.978	10.837	10.729	9.419	2.687	2.525	2.579	2.422	2.416	83 Bi
84 Po	93.112	89.231	89.781	92.383	79.279	76.851	16.935		13.635	13.083	16.244	15.741	13.445		13.817	13.338	11.129	11.014	9.662						84 Po
85 At	95.740	91.707	92.287	94.974	81.499	78.930	17.490		14.065		16.784	16.249	13.874		14.215		11.425	11.303							85 At
86 Rn	98.418	94.230	94.850	97.622	83.768	81.051	18.058		14.509		17.337	16.768	14.313		14.618		11.725	11.596							86 Rn
87 Fr	101.147	96.791	97.460	100.307	96.089	83.217	18.638		14.973		17.904	17.300	14.768		15.028	14.448	12.029	11.893							87 Fr
88 Ra	103.927	99.415	100.113	103.051	88.454	85.419	19.233	18.354	15.442	14.745	19.078	17.845	15.233	13.661	15.442	14.839	12.338	12.194	10.620						88 Ra
89 Ac	106.759	102.084	102.829	105.849	90.868	87.660	19.842		15.929		19.078	18.405	15.710		15.865		12.650	12.499							89 Ac
90 Th	109.649	104.813	105.591	108.699	93.334	89.938	20.470	19.503	16.423	15.640	19.692	18.979	16.199	14.507	16.300	15.621	12.967	12.807	11.117	3.491	3.145	3.332	2.996	2.986	90 Th
91 Pa	112.581	107.576	108.409	111.605	95.852	92.271	21.102	20.094	16.927	16.101	20.311	19.565	16.699	14.944	16.731	16.022	13.288	13.120	11.364		3.239		3.082	3.072	91 Pa
92 U	115.603	110.387	111.281	114.587	98.422	94.649	21.756	20.709	17.452	16.573	20.947	20.164	17.217	15.397	17.167	16.425	13.612	13.437	11.616	3.728	3.336	3.552	3.170	3.159	92 U
93 Np	118.619		113.725	118.057	100.781	96.844	22.417	21.336	17.986	17.058	21.596	20.781	17.747	15.874	17.614	16.837	13.942	13.757	11.887						93 Np
94 Pu	121.760		116.943	120.350	103.300	99.168	23.095	21.979	18.537	17.553	22.263	21.414	18.291	16.330	18.053	17.252	14.276	14.082	12.122						94 Pu
95 Am	124.876		120.350	123.960	105.949	101.607	23.793		19.103	18.060	22.944	22.061	18.849		18.525	17.673	14.615	14.409	12.381						95 Am
96 Cm	128.088		122.733	126.490	108.737	104.168	24.503				23.640	22.703	19.399		18.990	18.096	14.953	14.740							96 Cm
97 Bk	131.357		126.490	130.484	111.676	106.862	25.230				24.352	23.389	19.961		19.461	18.529	15.304	15.080							97 Bk
98 Cf	134.683		127.794	133.290	114.778	109.699	25.971				25.080	24.070	20.557		19.938	18.983	15.652	15.418							98 Cf

Appendix C: Vacuum System

All the passageways of electrons in a transmission electron microscope are kept at high vacuum. Figure C.1 shows a vacuum system of a typical analytical electron microscope with a field emission gun (JEM-2010F).

A sputter ion pump (SIP) of evacuation speed 15 l/s evacuates the space around the emitter exclusively to maintain stable emission. An SIP of evacuation speed 60 l/s evacuates the space of the acceleration tube exclusively to prevent discharges, and the space is kept at high vacuum (3×10^{-8} Pa). An intermediate chamber is set between the acceleration tube chamber and the column to create the differential pumping system. An SIP of evacuation speed 150 l/s evacuates both the specimen chamber and the column, and the vacuum of the specimen chamber is set to be higher than 3×10^{-5} Pa.

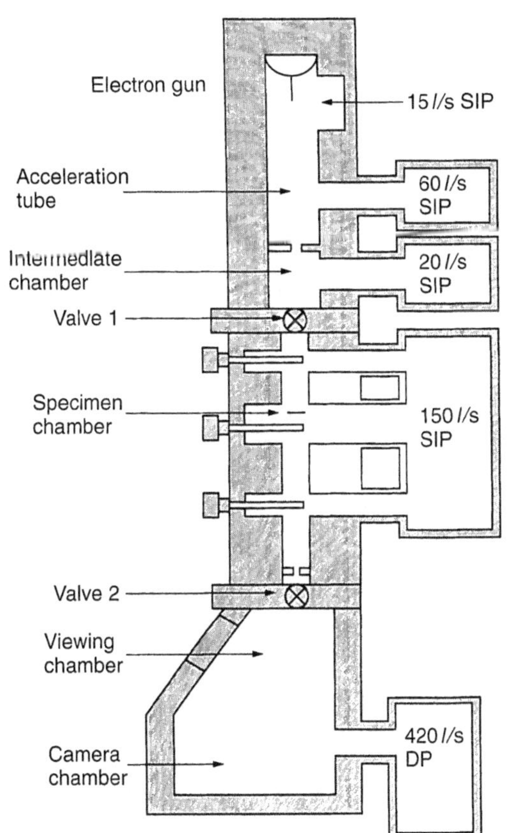

Fig. C.1. Vacuum system of a field emission transmission electron microscope (FE-TEM: JEM-2010F) microscope

Residual gas around the specimen generates specimen contamination, which obstructs observation and analysis. Therefore, the dry (oil-free) pumping system using SIP is employed to prevent hydrocarbon from invading.

The ability to bake the column and stage is added to the system. An anticontamination fin, which is cooled with liquid nitrogen, is set in around the specimen in the specimen chamber. An oil diffusion pump of evacuation speed 420 l/s evacuates the camera chamber, where much gas (mainly moisture) generates from photo-film. A small orifice is set in between the column and the camera chamber to create a differential pumping system. The back line of the diffusion pump is evacuated by a rotary pump. The principles and features of vacuum pumps employed in transmission electron microscopes are shown in Table C.1.

Table C.1. Principles and features of vacuum pumps employed in transmission electron microscopes.

Vacuum pump	Principles	Working vacuum range	Features and notes
Rotary pump (RP)	Pump sucks, compresses, and evacuates gas by rotation of a rotor in a chamber being kept tight and lubricious with oil.	Atmosphere-10^{-2} Pa	As the pump works from atmospheric pressure, it is employed for rough pumping for TEM. It is also used for back-line evacuation of an oil diffusion pump (DP) and a turbo molecular pump (TMP). Whenever the pump stops, the pump chamber should be set at atmospheric pressure to prevent the oil from flowing back.
Oil diffusion pump (DP)	Oil vapor jet is spouted from a nozzle by heating oil. Gas molecular is transferred away with the oil jet.	10^{-1}–10^{-8} Pa	As the pump works at a relatively lower vacuum level and has high evacuation speed, it is used to evacuate large-capacity camera chambers in which much gas is generated. Increased back-line pressure (decreasing vacuum) streams oil vapor back in the fore-line. Therefore, the back-line should be evacuated continuously with an RP.
Sputter ion pump (SIP)	Ions generated by magnetron discharges are sputtered on the surface of a titanium wall. The active molecules generated trap gas molecules, which are adsorbed on the inner wall.	10^{-2}–10^{-9} Pa	As the pump is oil-free, it is called the dry pumping system and is used to evacuate an electron gun chamber and a column. Because it adsorbs residual gas, it is unsuited for use where there is a lot of residual gas. It is better used to maintain high vacuum in the system. It is impossible to adsorb inert gas molecules, such as helium and argon. The pumping power is recovered by maintenance, where the pump is baked with evacuation by a DP.
Turbo molecular pump (TMP)	Gas molecules are evacuated by rotating a metal rotor fin at high speed.	10^{-2}–10^{-8} Pa	As the pump is working in the range from low to high vacuum and is oil-free, it is used to evacuate a column. To avoid vibrations, a magnetic buoyant-type rotor is employed. The back-line is evacuated by an RP.
Cryo pump (CP)	Gas molecules are adsorbed on the surface of a metal fin cooled with a coolant such as liquid nitrogen.	10^{-2}–10^{-13} Pa	Pump adsorbs all gas molecules including inert gas. It is possible to attain the ideal vacuum. An anticontamination fin installed in the specimen chamber is considered to be a kind of cryo pump.

Index

Table of Characteristic X-ray Energies (keV)

○ This table can be used to identify characteristic X-ray peaks

Notes:

1) This table is based on the values listed in Table B.2 (Appendix B).
2) Energies are listed up to 20 keV for K-M peaks; the lowest energy is given on the left.
3) In each series of K, L, and M peaks, three or fewer of the strongest peaks are listed from top to bottom.
4) Energy values are rounded to three decimal places.
5) When the difference between the energies of two peaks is less than 0.1 keV, the peaks are indicated together by $K_{\alpha1,2}$ with the energy value of the stronger peak $K_{\alpha1}$. Also, in cases when the energy of $K_{\beta1}$ differs from that of $K_{\alpha1}$ by less than 0.1 keV, the energy of $K_{\beta1}$ is given in parentheses.

From XES (X-ray Energy Spectrometry, Rolf Woldseth, published by Kevex Corporation, Burlingame, 1973.)

	IA	IIA	IIIA	IVA	VA	VIA	VIIA	VIII	
1	1 H Hydrogen								
2	3 Li Lithium $K_{\alpha1,2}$ 0.05	4 Be Beryllium $K_{\alpha1,2}$ 0.11							
3	11 Na Sodium $K_{\alpha1,2}$ 1.04 $K_{\beta1}$ (1.07)	12 Mg Magnesium $K_{\alpha1,2}$ 1.25 $K_{\beta1}$ (1.30)							
4	19 K Potassium $K_{\alpha1,2}$ 3.31 $K_{\beta1,3}$ 3.59	20 Ca Calcium $L_{\alpha1,2}$ 0.34 $K_{\alpha1,2}$ 3.69 $L_{\beta1}$ (0.35) $K_{\beta1,3}$ 4.01	21 Sc Scandium $L_{\alpha1,2}$ 0.40 $K_{\alpha1,2}$ 4.09 $L_{\beta1}$ (0.40) $K_{\beta1,3}$ 4.46	22 Ti Titanium $L_{\alpha1,2}$ 0.45 $K_{\alpha1,2}$ 4.51 $L_{\beta1}$ (0.46) $K_{\beta1,3}$ 4.93	23 V Vanadium $L_{\alpha1,2}$ 0.51 $K_{\alpha1,2}$ 4.95 $L_{\beta1}$ (0.52) $K_{\beta1,3}$ 5.43	24 Cr Chromium $L_{\alpha1,2}$ 0.57 $K_{\alpha1,2}$ 5.41 $L_{\beta1}$ (0.58) $K_{\beta1,3}$ 5.95	25 Mn Manganese $L_{\alpha1,2}$ 0.64 $K_{\alpha1,2}$ 5.90 $L_{\beta1}$ (0.65) $K_{\beta1,3}$ 6.49	26 Fe Iron $L_{\alpha1,2}$ 0.71 $K_{\alpha1,2}$ 6.40 $L_{\beta1}$ (0.72) $K_{\beta1,3}$ 7.06	27 Co Cobalt $L_{\alpha1,2}$ 0.78 $K_{\alpha1,2}$ 6.93 $L_{\beta1}$ (0.79) $K_{\beta1,3}$ 7.65
5	37 Rb Rubidium $L_{\alpha1,2}$ 1.69 $K_{\alpha1,2}$ 13.39 $L_{\beta1}$ (1.75) $K_{\beta1,3}$ 14.96 $K_{\beta2}$ 15.18	38 Sr Strontium $L_{\alpha1,2}$ 1.81 $K_{\alpha1,2}$ 14.16 $L_{\beta1}$ (1.87) $K_{\beta1,3}$ 15.83 $K_{\beta2}$ 16.08	39 Y Yttrium $L_{\alpha1,2}$ 1.92 $K_{\alpha1,2}$ 14.96 $L_{\beta1}$ (2.00) $K_{\beta1,3}$ 16.74 $K_{\beta2}$ 17.01	40 Zr Zirconium $L_{\alpha1,2}$ 2.04 $K_{\alpha1,2}$ 15.77 $L_{\beta1}$ (2.12) $K_{\beta1,3}$ 17.67 $L_{\beta2}$ 2.22 $K_{\beta2}$ 17.97	41 Nb Niobium $L_{\alpha1,2}$ 2.17 $K_{\alpha1,2}$ 16.61 $L_{\beta1}$ (2.26) $K_{\beta1,3}$ 18.62 $L_{\beta2}$ 2.37 $K_{\beta2}$ 18.95	42 Mo Molybdenum $L_{\alpha1,2}$ 2.29 $K_{\alpha1}$ 17.48 $L_{\beta1}$ 2.39 $K_{\alpha2}$ 17.37 $L_{\beta2}$ 2.52 $K_{\beta1,3}$ 19.61	43 Tc Technetium $L_{\alpha1,2}$ 2.42 $K_{\alpha1}$ 18.36 $L_{\beta1}$ 2.54 $K_{\alpha2}$ 18.25 $L_{\beta2}$ —	44 Ru Ruthenium $L_{\alpha1,2}$ 2.56 $K_{\alpha1}$ 19.28 $L_{\beta1}$ 2.68 $K_{\alpha2}$ 19.15 $L_{\beta2}$ 2.84	45 Rh Rhodium $L_{\alpha1,2}$ 2.70 $L_{\beta1}$ 2.83 $L_{\beta2}$ 3.00
6	55 Cs Cesium $L_{\alpha1,2}$ 4.29 $L_{\beta1}$ 4.62 $L_{\beta2}$ 4.94	56 Ba Barium $L_{\alpha1,2}$ 4.47 $L_{\beta1}$ 4.83 $L_{\beta2}$ 5.16	57 La ~ 71 Lu	72 Hf Hafnium M_{β} 1.70 $L_{\alpha1}$ 7.90 $L_{\beta1}$ 9.02 $L_{\beta2}$ 9.35	73 Ta Tantalum M_{β} 1.77 $L_{\alpha1}$ 8.15 $L_{\beta1}$ 9.34 $L_{\beta2}$ 9.65	74 W Tungsten $M_{\alpha1,2}$ 1.78 $L_{\alpha1}$ 8.40 M_{β} (1.84) $L_{\beta1}$ 9.67 $L_{\beta2}$ 9.96	75 Re Rhenium $M_{\alpha1,2}$ — $L_{\alpha1}$ 8.65 M_{β} 1.91 $L_{\beta1}$ 10.01 $L_{\beta2}$ 10.27	76 Os Osmium $M_{\alpha1,2}$ — $L_{\alpha1}$ 8.91 M_{β} 1.98 $L_{\beta1}$ 10.35 $L_{\beta2}$ 10.60	77 Ir Iridium $M_{\alpha1,2}$ 1.98 $L_{\alpha1}$ 9.17 M_{β} (2.05) $L_{\beta1}$ 10.71 $L_{\beta2}$ 10.92
7	87 Fr Francium $M_{\alpha1,2}$ — $L_{\alpha1}$ 12.03 M_{β} — $L_{\beta1}$ 14.77 $L_{\beta2}$ 14.45	88 Ra Radium $M_{\alpha1,2}$ — $L_{\alpha1}$ 12.34 M_{β} $L_{\beta1}$ 15.23 $L_{\beta2}$ 14.84	89 Ac ~ 103 Lr						

57 La Lanthanum	58 Ce Cerium	59 Pr Praseodymium	60 Nd Neodymium	61 Pm Promethium	62 Sm Samarium
M_{β} 0.85 $L_{\alpha1,2}$ 4.65 $L_{\beta1}$ 5.04 $L_{\beta2}$ 5.38	M_{β} 0.90 $L_{\alpha1,2}$ 4.84 $L_{\beta1}$ 5.26 $L_{\beta2}$ 5.61	M_{β} 0.95 $L_{\alpha1,2}$ 5.03 $L_{\beta1}$ 5.49 $L_{\beta2}$ 5.85	M_{β} 1.00 $L_{\alpha1,2}$ 5.23 $L_{\beta1}$ 5.72 $L_{\beta2}$ 6.09	M_{β} — $L_{\alpha1,2}$ 5.43 $L_{\beta1}$ 5.96 $L_{\beta2}$ 6.34	M_{β} 1.10 $L_{\alpha1,2}$ 5.64 $L_{\beta1}$ 6.21 $L_{\beta2}$ 6.59
89 Ac **Actinium**	**90 Th** **Thorium**	**91 Pa** **Protactinium**	**92 U** **Uranium**	**93 Np** **Neptunium**	**94 Pu** **Plutonium**
$M_{\alpha1,2}$ — $L_{\alpha1}$ 12.65 M_{β} — $L_{\beta1}$ 15.71 $L_{\beta2}$ —	$M_{\alpha1,2}$ 3.00 $L_{\alpha1}$ 12.97 M_{β} 3.15 $L_{\beta1}$ 16.20 $L_{\beta2}$ 15.62	$M_{\alpha1,2}$ 3.08 $L_{\alpha1}$ 13.29 M_{β} 3.24 $L_{\beta1}$ 16.70 $L_{\beta2}$ 16.02	$M_{\alpha1,2}$ 3.17 $L_{\alpha1}$ 13.61 M_{β} 3.34 $L_{\beta1}$ 17.22 $L_{\beta2}$ 16.43	$M_{\alpha1,2}$ — $L_{\alpha1}$ 13.94 M_{β} — $L_{\beta1}$ 17.75 $L_{\beta2}$ 16.84	$L_{\alpha1}$ 14.28 $L_{\beta1}$ 18.29 $L_{\beta2}$ 17.25

	I B	II B	III B	IV B	V B	VI B	VII B	0
								2 He Helium
			5 B Boron $K_{\alpha1,2}$ 0.19	6 C Carbon $K_{\alpha1,2}$ 0.28	7 N Nitrogen $K_{\alpha1,2}$ 0.39	8 O Oxygen $K_{\alpha1,2}$ 0.53	9 F Fluorine $K_{\alpha1,2}$ 0.68	10 Ne Neon $K_{\alpha1,2}$ 0.85
			13 Al Aluminum $K_{\alpha1,2}$ 1.49 $K_{\beta1}$ (1.55)	14 Si Silicon $K_{\alpha1,2}$ 1.74 $K_{\beta1}$ (1.83)	15 P Phosphorus $K_{\alpha1,2}$ 2.01 $K_{\beta1}$ 2.14	16 S Sulfur $K_{\alpha1,2}$ 2.31 $K_{\beta1}$ 2.46	17 Cl Chorine $K_{\alpha1,2}$ 2.62 $K_{\beta1}$ —	18 Ar Argon $K_{\alpha1,2}$ 2.96 $K_{\beta1,3}$ 3.19
28 Ni Nickel $L_{\alpha1,2}$ 0.85 $K_{\alpha1,2}$ 7.48 $L_{\beta1}$ (0.87) $K_{\beta1,3}$ 8.26	29 Cu Copper $L_{\alpha1,2}$ 0.93 $K_{\alpha1,2}$ 8.05 $L_{\beta1}$ (0.95) $K_{\beta1,3}$ 8.90	30 Zn Zinc $L_{\alpha1,2}$ 1.01 $K_{\alpha1,2}$ 8.64 $L_{\beta1}$ (1.03) $K_{\beta1,3}$ 9.57	31 Ga Gallium $L_{\alpha1,2}$ 1.10 $K_{\alpha1,2}$ 9.25 $L_{\beta1}$ (1.13) $K_{\beta1,3}$ 10.26 $K_{\beta2}$ 10.37	32 Ge Germanium $L_{\alpha1,2}$ 1.19 $K_{\alpha1,2}$ 9.89 $L_{\beta1}$ (1.22) $K_{\beta1,3}$ 10.98 $K_{\beta2}$ 11.10	33 As Arsenic $L_{\alpha1,2}$ 1.28 $K_{\alpha1,2}$ 10.54 $L_{\beta1}$ (1.32) $K_{\beta1,3}$ 11.72 $K_{\beta2}$ 11.86	34 Se Selenium $L_{\alpha1,2}$ 1.38 $K_{\alpha1,2}$ 11.23 $L_{\beta1}$ (1.42) $K_{\beta1,3}$ 12.49 $K_{\beta2}$ 12.65	35 Br Bromine $L_{\alpha1,2}$ 1.48 $K_{\alpha1,2}$ 11.92 $L_{\beta1}$ (1.53) $K_{\beta1,3}$ 13.29 $K_{\beta2}$ 13.47	36 Kr Krypton $L_{\alpha1,2}$ 1.59 $K_{\alpha1,2}$ 12.65 $L_{\beta1}$ (1.64) $K_{\beta1,3}$ 14.11 $K_{\beta2}$ 14.31
46 Pd Palladium $L_{\alpha1,2}$ 2.84 $L_{\beta1}$ 2.99 $L_{\beta2}$ 3.17	47 Ag Silver $L_{\alpha1,2}$ 2.98 $L_{\beta1}$ 3.15 $L_{\beta2}$ 3.35	48 Cd Cadmium $L_{\alpha1,2}$ 3.13 $L_{\beta1}$ 3.32 $L_{\beta2}$ 3.53	49 In Indium $L_{\alpha1,2}$ 3.29 $L_{\beta1}$ 3.49 $L_{\beta2}$ 3.71	50 Sn Tin $L_{\alpha1,2}$ 3.44 $L_{\beta1}$ 3.66 $L_{\beta2}$ 3.90	51 Sb Antimony $L_{\alpha1,2}$ 3.60 $L_{\beta1}$ 3.84 $L_{\beta2}$ 4.10	52 Te Tellurium $L_{\alpha1,2}$ 3.77 $L_{\beta1}$ 4.03 $L_{\beta2}$ 4.30	53 I Iodine $L_{\alpha1,2}$ 3.94 $L_{\beta1}$ 4.22 $L_{\beta2}$ 4.51	54 Xe Xenon $L_{\alpha1,2}$ 4.11 $L_{\beta1}$ — $L_{\beta2}$ —
78 Pt Platinum $M_{\alpha1,2}$ 2.05 $L_{\alpha1}$ 9.44 M_{β} (2.13) $L_{\beta1}$ 11.07 $L_{\beta2}$ 11.25	79 Au Gold $M_{\alpha1,2}$ 2.12 $L_{\alpha1}$ 9.71 M_{β} (2.20) $L_{\beta1}$ 11.44 $L_{\beta2}$ 11.58	80 Hg Mercury $M_{\alpha1,2}$ — $L_{\alpha1}$ 9.99 M_{β} 2.28 $L_{\beta1}$ 11.82 $L_{\beta2}$ 11.92	81 Tl Thallium $M_{\alpha1,2}$ 2.27 $L_{\alpha1}$ 10.27 M_{β} (2.36) $L_{\beta1}$ 12.21 $L_{\beta2}$ (12.27)	82 Pb Lead $M_{\alpha1,2}$ 2.35 $L_{\alpha1}$ 10.55 M_{β} (2.44) $L_{\beta1}$ 12.61 $L_{\beta2}$ (12.62)	83 Bi Bismuth $M_{\alpha1,2}$ 2.42 $L_{\alpha1}$ 10.84 M_{β} 2.52 $L_{\beta1}$ 13.02 $L_{\beta2}$ (12.98)	84 Po Polonium $M_{\alpha1,2}$ — $L_{\alpha1}$ 11.13 M_{β} — $L_{\beta1}$ 13.45 $L_{\beta2}$ 13.34	85 At Astatine $M_{\alpha1,2}$ — $L_{\alpha1}$ 11.43 M_{β} — $L_{\beta1}$ 13.87 $L_{\beta2}$ —	86 Rn Radon $M_{\alpha1,2}$ — $L_{\alpha1}$ 11.73 M_{β} — $L_{\beta1}$ 14.31 $L_{\beta2}$ —

Symbol

Atomic number → 8 O

Oxygen ← Name

$K_{\alpha1,2}$ ← Characteristic X-ray

0.53 ← Energy (keV)

Relative intensities of main characteristic X-ray peaks

$K\alpha_1=100$	$L\alpha_1=100$	$M\alpha_{1,2}=100$
$K\alpha_2=50$	$L\alpha_2=10$	$M\beta=\sim60$
$K\alpha_{1,2}=150$	$L\beta_1=50$	
$K\beta_1=15\text{-}30$	$L\beta_2=20$	
$K\beta_2=1\text{-}10$	$L\beta_3=1\text{-}6$	
$K\beta_3=6\text{-}15$	$L\beta_4=3\text{-}5$	

63 Eu Europium M_{β} 1.15 $L_{\alpha1}$ 5.85 $L_{\beta1}$ 6.46 $L_{\beta2}$ 6.84	64 Gd Gadolinium M_{β} 1.21 $L_{\alpha1}$ 6.06 $L_{\beta1}$ 6.71 $L_{\beta2}$ 7.10	65 Tb Terbium M_{β} 1.27 $L_{\alpha1}$ 6.27 $L_{\beta1}$ 6.98 $L_{\beta2}$ 7.37	66 Dy Dysprosium M_{β} 1.33 $L_{\alpha1}$ 6.49 $L_{\beta1}$ 7.25 $L_{\beta2}$ 7.63	67 Ho Holmium M_{β} 1.38 $L_{\alpha1}$ 6.72 $L_{\beta1}$ 7.52 $L_{\beta2}$ 7.91	68 Er Erbium M_{β} 1.44 $L_{\alpha1}$ 6.95 $L_{\beta1}$ 7.81 $L_{\beta2}$ 8.19	69 Tm Thulium M_{β} 1.50 $L_{\alpha1}$ 7.18 $L_{\beta1}$ 8.10 $L_{\beta2}$ 8.47	70 Yb Ytterbium M_{β} 1.57 $L_{\alpha1}$ 7.41 $L_{\beta1}$ 8.40 $L_{\beta2}$ 8.76	71 Lu Lutetium M_{β} 1.63 $L_{\alpha1}$ 7.65 $L_{\beta1}$ 8.71 $L_{\beta2}$ 9.05
95 Am Americium $L_{\alpha1}$ 14.62 $L_{\beta1}$ 18.85 $L_{\beta2}$ 17.67	96 Cm Curium $L_{\alpha1}$ 14.95 $L_{\beta1}$ 19.40 $L_{\beta2}$ 18.10	97 Bk Berkelium $L_{\alpha1}$ 15.30 $L_{\beta1}$ 19.96 $L_{\beta2}$ 18.53	98 Cf Californium $L_{\alpha1}$ 15.65 $L_{\beta2}$ 18.98 $L_{\alpha2}$ 15.42	99 Es Einsteinium	100 Fm Fermium	101 Md Mendelevium	102 No Nobelium	103 Lr Lawrencium

GPSR Compliance

The European Union's (EU) General Product Safety Regulation (GPSR) is a set of rules that requires consumer products to be safe and our obligations to ensure this.

If you have any concerns about our products, you can contact us on ProductSafety@springernature.com

In case Publisher is established outside the EU, the EU authorized representative is:

Springer Nature Customer Service Center GmbH
Europaplatz 3
69115 Heidelberg, Germany

Batch number: 09625567

Printed by Printforce, the Netherlands